SCIENCE & MUSIC

Baghdad Museum

Fig. 1. Dancing to the lyre, and to the clapping of curved sticks (about 2700 B.C.) From a royal tomb, Ur.

University Museum, Philadelphia

Fig. 2. Music from a four-stringed, bow-shaped harp (on right) and other instruments (about 2700 B.C.). From the tomb of Queen Shubad, Ur.

SUMERIAN MUSIC OF 4600 YEARS AGO

SCIENCE & MUSIC

by
SIR JAMES JEANS

DOVER PUBLICATIONS, INC.
NEW YORK

Published in Canada by General Publishing Company, Ltd., 30 Lesmill Road, Don Mills, Toronto, Ontario.

Published in the United Kingdom by Constable and Company, Ltd.

This Dover edition, first published in 1968, is an unabridged republication of the work originally published in 1937, and is reprinted by permission of the Cambridge University Press.

International Standard Book Number: 0-486-61964-8

Library of Congress Catalog Card Number: 68-24652

Manufactured in the United States of America
Dover Publications, Inc.
180 Varick Street
New York, N. Y. 10014

TO
MY WIFE

CONTENTS

PLATES

The frontispiece is taken from *The Music of the Sumerians, Babylonians and Assyrians* (Cambridge, 1937), by courtesy of the author, Canon F. W. Galpin; Plates I, II, III, VIII and IX are from photographs kindly supplied by Professor Dayton C. Miller; Plate IV is taken from an article in the *Journal of the Acoustical Society of America*, by courtesy of the Editor and of Mr Hart; Plates V and VI are reproduced from a paper in the *Philosophical Transactions of the Royal Society*, 1934, by kind permission of the author, Dr G. J. Richards, and of the Society. Plate VII first appeared in the *Journal de Physique*, and is here reproduced by courtesy of La Société Française de Physique, and of Messrs Edward Arnold, who supplied the blocks.

PREFACE

Much has been added to our scientific knowledge of musical sound, since Helmholtz published his great work *Tonempfind-ungen* in 1862. The new knowledge has been often and well described, but mostly by scientists writing for scientists in the technical language of science.

In the present book I have tried to describe the main outlines of such parts of science, both old and new, as are specially related to the questions and problems of music, assuming no previous knowledge either of science or of mathematics on the part of the reader. My aim has been to convey precise information in a simple non-technical way, and I hope the subject-matter I have selected may interest the amateur, as well as the serious student, of music.

I need hardly say that I am indebted to many friends and books. A considerable fraction of my book is merely Helmholtz modernised and rewritten in simple language. Another considerable fraction is drawn from the wealth of material provided in the notes added to Helmholtz's book by his English translator, A. J. Ellis. On the less technical side, I have borrowed largely from Dayton C. Miller's book *The Science of Musical Sounds* (The Macmillan Company, 1934), and am especially indebted to the author for permission to reproduce eleven excellent photographs of sound-curves. Among other sources from which I have drawn largely, and found especially valuable, I ought to mention:

Sound by Lord Rayleigh (2 vols. Macmillan & Co.);

Sound by F. R. Watson (John Wiley, 1935);

A Text-book of Sound by A. B. Wood (Bell, 1932);

Hearing in Man and Animals by R. T. Beatty (Bell, 1932);

Physical Society of London: Report of a Discussion on Audition (1931);

Physical Society of London: Reports on Progress in Physics. Vol. II, 1935, and Vol. III, 1937;

Modern Acoustics by A. H. Davis (Bell, 1934);

The Acoustics of Orchestral Instruments and of the Organ by E. G. Richardson (Arnold, 1929);

The Acoustics of Buildings by A. H. Davis and G. W. C. Kaye (Bell, 1932);

Collected Papers on Acoustics by W. C. Sabine (Harvard University Press, 1927);

as well as innumerable papers in technical and scientific journals.

On the personal side, I am especially indebted to my wife, to Henry Willis and to Philip Pfaff, Mus.Bac.

J. H. JEANS

Dorking
June 1937

SCIENCE & MUSIC

CHAPTER I

INTRODUCTION

The Coming of Music

The lantern of science, throwing its light down the long corridors of time, enables us to trace out the gradual evolution of terrestrial life. Far away in the dim distances of the remote past we see it emerging from lowly beginnings —possibly single-cell organisms on the sea shore—and gradually increasing in complexity until it culminates in the higher mammals of to-day, and in man, the most complicated form of life which has so far emerged from the workshop of nature. And as living beings become more complex, they acquire an ever more intricate battery of sense-organs which help them to find their way about the world, to escape danger, to capture their food and avoid being themselves captured as food.

One of these is of special interest to musicians, for out of it has developed our present organ of hearing. Sunk into the skin of a fish, and running the whole length of its body, from head to tail on either side, there is a line of pits or depressions. Under these lies an organ known as the "lateral-line" organ. This is believed to register differences of pressure in the water, which will acquaint the fish with the currents and eddies in which he is swimming, and may also warn him of the proximity of other fish, especially of large fish of hostile intentions.

Even the most primitive fishes seem to have possessed a simple organ of this kind. Gradually the depression nearest to the head developed into something far more intricate, namely the hard bony structure known as the "labyrinth", which is found in all vertebrates, including ourselves. It consists of hollow tubes filled with fluid, and the main part of it is shaped so as to form three (or in rare cases only one or two) semicircular canals, lying in directions mutually at right angles to one another, as on the right of fig. 1.

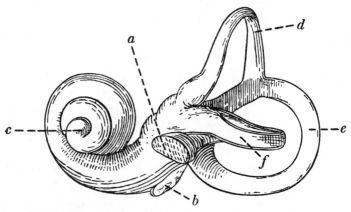

Fig. 1. The labyrinth of the left human ear (magnified about 5 times). The three semicircular canals are on the right (*d, e, f*) and the cochlea on the left (*c*). *a* is the oval window to which the ear-drum transmits its vibrations; *b* is the round window, the function of which is explained below (p. 246).

When an animal turns its head or the upper part of its body, the fluid in the semicircular canals lags behind, because of its inertia, and so rubs over a set of paint-brushes of fine hairs, one in each canal; the bending of these hairs sends a series of nerve-impulses to the brain, which inform it of the change of direction and initiate a set of reflex actions to balance the change. Human beings are seldom conscious that they possess such organs, although it is by their help that we regain our balance after a sudden slip. They are also responsible for the giddiness we feel after spinning round too often or too rapidly, and for part at least of the even less agreeable sensations we experience when we are on a small ship in a turbulent sea.

A simple equipment of this kind would be adequate for primaeval fish, which lived entirely in the water, but would soon prove inadequate under new conditions which were to come. For the geologists tell us of a period of great drought occurring some 300 million years ago, when seas, lakes and marshes were all drying up. It must have been an anxious time for the fishes, many of which would desert their pools and shallows, and flop across dry land in the hope of finding new water. Clearly the more amphibious they could become, the greater was their chance of survival. In time some of the survivors became pure land-animals—our own ancestry. Organs for registering differences of pressure in water would be of little use to them now. What they needed was an organ to register minute differences of pressure in air, for these were associated with sounds which might indicate the presence of food or of danger, of friends or of enemies.

Gradually the required new organ seems to have developed out of the old. The story of the change provides one of the most fascinating—and, one is almost tempted to say, most incredible—chapters in the evolutionary record. A small area of the bony structure of the labyrinth became thinned down into a yielding membrane of mere skin, thin and soft enough to transmit variations of pressure from the air outside to the fluid within. At the same time, the labyrinth itself grew in size and increased in complexity. That of the frog shews a small bulge, which, as we proceed farther upwards in the scale of life, gradually develops into the cochlea, which forms the essential part of the ear of vertebrates. The external appearance of this wonderfully intricate piece of apparatus is shewn in fig. 1 on p. 2; its interior is described later (p. 246). For the moment we can only compare it to the case, the soundboard and the strings of a pianoforte of many strings— about 3000 in birds, 16,000 in cats and 24,000 in man— all compressed to the dimensions of less than a pea. It enables its possessor not only to hear sounds, but also to analyse them into their constituent tones. This power of analysis must obviously have had a great "survival value" for primitive life, since sounds which have been analysed can be remembered, and those which have once been found to be associated with danger can be promptly acted upon when heard again—just as we do with the motor-horn in our less primitive life of to-day.

In some such way as this, the human race became possessed of its ears. At first they would merely be helps in the struggle for existence. But we can imagine primitive man one day discovering in them an interest and a value of another kind; we can imagine him finding that the hearing of some simple sound, perhaps the twang of his bowstring or the blowing of the wind over a broken reed, was a pleasure in itself. On that day music was born, and from that day to this innumerable workers of many ages and of many peoples have been trying to discover new sounds of a pleasure-giving kind, and to master the art of blending and weaving these together so as to give the maximum of enjoyment, with the result that music of one kind or another now figures largely in the lives of most civilised beings.

The Sense of Hearing

As life slowly climbed the long ladder of evolution, one sense after another arrived and developed. Hearing was the last to arrive, and the last to attain a state bordering on perfection. When it reached this state, the other senses were already highly developed, and one, the sense of seeing, had already attained too much importance to be displaced. For most animals seeing must always have been more important than hearing, and whether we think in terms of our pleasure or of our well-being we must admit that the same is true for us to-day; we would sooner lose any of our other senses than that of sight. Throughout most of our waking life, we are seeing and hearing at the same time, and our sensations of sight are usually far more intense than those of sound. And as we obtain more pleasure through our eyes than through our ears, we have acquired the habit of giving the greater part of our attention to what we see, leaving a mere fraction for what we hear. Not only so, but hearing and seeing do not blend well; they rather compete—in an unequal competition in which seeing usually wins. In the opera house, many of us miss much of the music through watching the acting too intently. Only when the distraction of sight is removed can our minds give full attention to what we hear. Our appreciation of sound then becomes far keener and more critical. This is why blind people so often become exceptionally good musicians, and why many people who are not blind find it well to listen to the radio with the room darkened, and to close their eyes in the concert room, resisting the temptation to watch the fingers of the pianist, or the mouth of the prima donna.

The Human Ear

The visible part of the ear consists of an external shell, the relics of an earlier sound-collector, with an aperture—the "meatus" or auditory canal—somewhere in its lower half. At the far end of this canal, approximately an inch inside the head, is a small delicate membrane of skin, only about three thousandths of an inch in thickness. It is oval in shape, being about a third of an inch in height, and two-fifths of an inch in width. It is stretched tightly over a hard frame of bone, much as the skin of a drum is stretched over a hard frame of wood; because of this it is known as the "membrana tympani", or ear-drum (see fig. 2, p. 9 below).

Sound reaches our ears in the form of waves which have travelled through the surrounding air, much as waves travel over the surface of a sea or river; some of these waves travel down the inch-long backwater formed by the auditory canal, and finally encounter the ear-drum, which forms a barrier at the far end.

When water-waves are stopped by a barrier, the pressure they exert on it varies with the rise and fall of the waves, and the variations of pressure may shake it into motion. We may often feel a sea-wall tremble under the pounding of the waves, and a delicate seismograph many miles inland will record the impact of sea-waves on a rocky coast. In the same way, sound-waves in air exert a varying pressure on our ear-drums which may set them into motion. But there is one essential difference. The sea-wall may be shaken to pieces in a few years, but the ear-drum has the capacity of continually renovating itself, and so keeping its efficiency almost unimpaired. Even if it is completely shattered by the intense noise of an explosion or a gun-blast, it will renew itself in a few weeks.

Our ear-drums are sensitive to an almost inconceivable degree. The tiniest ripple in the air sets them into motion; under favourable conditions a sound-wave of such feeble intensity that the air is displaced only through a ten-thousand-millionth part of an inch will send an audible sound to the brain. The change of pressure produced by such a sound-wave is less than a ten-thousand-millionth part of the whole pressure of the atmosphere, so that the human ear is incomparably more sensitive than any baro-meter which has ever been constructed. The ordinary barometer will record the lowering of atmospheric pressure which we experience as we walk upstairs in our house, or climb a few feet up the mountain-side, but the change of pressure just mentioned is that produced by an ascent of only a 30,000th of an inch. The feeblest nodding of our head changes the pressure on our ear-drums by more than is necessary to set them into motion, and if we do not hear a musical sound, it is only because we cannot nod our heads with sufficient rapidity. For, although our ear-drums are very sensitive to minute changes of pressure, it is only when these changes are repeated in rapid succession that messages are passed on to the brain. We shall see later why this is.

Immediately behind the ear-drum lies a chain of small bones, known as "ossicles". The first of these is in contact

8   INTRODUCTION

with the ear-drum, while the last presses against the "oval window" of the labyrinth, the thin yielding membrane of skin already described (*a* in fig. 1). The ossicles transmit the motion of the ear-drum to this oval window much as a bell-wire transmits a pull from a bell-rope to a bell. The oval window passes the motion on to the fluid inside the labyrinth, and in this way it reaches the cochlea—the miniature pianoforte which has already been mentioned. The workings of the cochlea are not yet fully understood, but we know that out of it emerges a bundle of nerves, and that when the ear-drum is set into vibration, minute currents of electricity pass through these nerves to the brain, and produce in it sensations which keep it informed as to the vibratory motions of the ear-drum.

The Process of Hearing

To obtain a more precise picture of the process of hearing, let us imagine that we are listening to an ordinary telephone conversation.

The essentials are shewn diagrammatically in fig. 2. The ear is on the right, and is open to the air as far as the ear-drum *d*. The telephone is on the left, and is open to the air as far as a metal diaphragm *D*. We at once notice a sort of symmetry between the two instruments, the solid cartilage and bone of the ear corresponding to the vulcanite framework of the telephone, while the ear-drum *d* corresponds to the diaphragm *D* of the telephone.

This diaphragm, like the ear-drum, has a complex piece of apparatus behind it, but out of this only a single pair of wires emerges. This is the telephone line, which may have its other end hundreds of miles away. Its function is to bring into the telephone electric currents which represent sound produced at its other end. The telephone transforms these currents into motions of the diaphragm, and so acts in just the opposite way to the ear, which transforms motions of the ear-drum into electric currents.

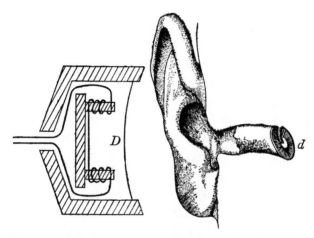

Fig. 2. Diagrammatic representation of the process of hearing. The action of the ear is somewhat like that of a telephone, but reversed. The telephone transforms the variations of an electric current into the vibrations of a diaphragm D, while the ear transforms the vibrations of the ear-drum d into electric currents which transmit sensations to the brain.

The apparatus behind the diaphragm of the telephone consists primarily of a magnet of the rather special kind known as a "polarising" magnet. Unlike the familiar horse-shoe magnet, this is not made of magnetised steel throughout, but has two projecting ends of soft iron. The telephone line makes several turns round each of these. Now a well-known law of physics tells us that a piece of soft iron which is encircled by an electric current becomes a temporary magnet, and so attracts any steel or iron which may be in its proximity, for so long as the current is flowing. In our diagram the magnet attracts the diaphragm D all the time, but when an electric current is flowing through the telephone line, the two pieces of soft iron form an additional magnet, and so give an extra pull to the diaphragm.

When we are listening to a telephone conversation, the current in the telephone line is not of unvarying strength; it continually waxes and wanes. As a result of this varying current, the diaphragm D experiences an extra pull which also waxes and wanes; it is pulled at one moment weakly, at another forcibly, at still another not at all, and so is kept continually in motion. Each time it moves a bit to the right, the air between it and the ear-drum is pushed a bit to the right, so that the ear-drum itself is pushed a bit to the right. Conversely, when the diaphragm moves to the left, the air is sucked outwards and draws the ear-drum to the left. In brief, we may say that the motion of the ear-drum reproduces that of the diaphragm, and this in turn reproduces the changes in the strength of the current in the wire.

In the ear exactly the converse process is taking place. While the telephone receiver is transforming the variations of electric currents in the wire into a mechanical motion of the diaphragm, the ear is transforming the resulting mechanical motion of the ear-drum into electric currents of varying intensity in the nerves which lead to the brain, and these currents result in our hearing the sound. We shall discuss the mechanism of the transformation later (p. 245). For the moment we return to our telephone.

Sound-Curves

The current flowing in the telephone wire at any instant can be measured with simple electrical instruments, and its changes can be represented on a chart, like that on which the recording barometer exhibits changes in the pressure of the atmosphere. In such a chart a roll of paper is drawn horizontally and at a uniform rate under a pen, which is

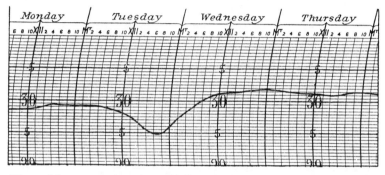

Fig. 3. A barometer chart. The horizontal scale indicates the passage of time, while the vertical scale shews the height of the barometer at each of the instants represented on the horizontal scale. We see, for instance, that at noon on Tuesday the barometer stood at 29·8 inches.

connected with an ordinary barometer. As the height of the barometer changes the pen moves up and down, and so draws a curve (see fig. 3) which records the variations of pressure.

We can easily imagine a similar chart in which the passage of time is again represented by motion in a horizontal direction, while vertical height no longer represents the height of a barometer, but the strength of the current flowing in the telephone wire. The units in which time is measured will no longer be whole days, but perhaps hundredths of a second, while the units of current may be anything suitable, but will certainly be something quite small.

We shall again be able to represent the fluctuations in the current by a curve of the same general nature as that of the barometer record—such a curve, let us say, as is shewn in fig. 4.

Seconds

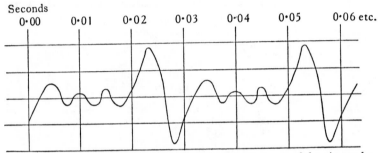

Fig. 4. A current chart. Just as variations of the pressure of the air can be represented by a curve in the way shewn in fig. 3, so the variations of the current in a wire can be represented by a curve such as that shewn above.

The motions of the diaphragm D of the telephone, or of the ear-drum, can also be represented on an exactly similar chart, except that the vertical units will now represent small units of length—perhaps millionths of an inch.

Thus we see that the current which conveys sound, the motion of the diaphragm which transmits this sound to the ear, and the motion of the ear-drum itself, can all be represented by curves of the kind shewn in fig. 4. And as the motion of the ear-drum follows that of the diaphragm, while this in turn follows the changes in the current, these curves will all be similar in shape. Each of them represents a certain sound, or succession of sounds. Further, as we know that all sounds, whether produced by nature or

PLATE I

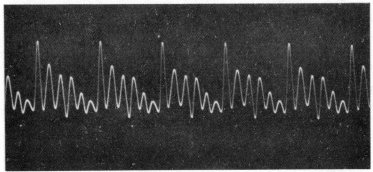

Dayton C. Miller

Fig. 1. The sound-curve of the vowel $\bar{a}$ in *father*, intoned by a bass voice at pitch F. The dots below the curve indicate intervals of $\frac{1}{100}$ second.

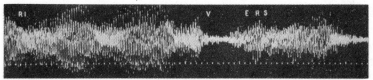

Dayton C. Miller

Fig. 2. The sound-curve of a gramophone record of a baritone voice singing the word *rivers* to the accompaniment of an orchestra. The dots below the curve again indicate intervals of $\frac{1}{100}$ second, so that the curve is much more compressed than that shewn in fig. 1.

TYPICAL SOUND-CURVES

by art, can be transmitted by telephone, it follows that all sounds can be represented by such a curve—a cough or a sneeze, the voice of a friend, or an orchestra playing a symphony. Such curves are now easily recorded by various instruments, the cathode-ray oscillograph in particular. Many photographs of curves taken by this and other forms of oscillograph will be found in the present book. The curves shewn on Plate I may be regarded as typical of many; they are the curves of a bass voice singing the vowel *ā* in *father*, and of a baritone voice singing the word *rivers* to an orchestral accompaniment.

Before a symphony can be played by an orchestra there must be collaboration of many parties—a composer, the makers and the players of many instruments and the conductor of the orchestra. All are, or have been, at work to produce—just a curve. If they have done their work well, the sound that this curve represents will be both pleasing to our ears and interesting to our minds. The composer, in writing his score, has given a first rough indication of the curve he desires—he has, so to speak, specified its main ingredients, and the instants at which they are to join the general mêlée. It is the business of the instrument-maker and the players to see that these ingredients are of good quality, while the function of the conductor is to see that they join in at the right moments and in the right proportions. All the art, all the mannerisms, all the successes and failures of these many workers are embodied in the one single curve. This curve *is* the symphony—neither more nor less, and the symphony will sound noble or tawdry, musical or harsh, refined or vulgar, according to the quality of this curve.

When a gramophone record is made of the performance of the symphony, this curve is preserved in a tangible form; it is nothing more or less than the shape of the uneven rim of the groove in which the gramophone needle runs when the record is played. On playing the record, we transform the curve into the music it represents. As the point of the needle is dragged along the groove—or, more accurately, as the groove is dragged under the point of the needle—the unevennesses in the walls of this groove make the needle move continually to the right and left. The blunt end of the needle transmits this motion to a mica diaphragm, which in turn imparts it to the surrounding air. The air then conveys the motion to our ear-drums, much as it was conveyed from the diaphragm of the telephone in fig. 2, except that the greater distance of our ears introduces a complication which we shall discuss later. In this way our ear-drums are made to vibrate in response to the curve which forms the rim of the groove of the record, and our brains are made conscious of the music that the curve represents.

The Transmission of Sound

If we had perfect materials at our disposal, ideal in quality and unlimited in quantity, it would be a simple matter to arrange that the curve received by our ears should be exactly identical with that which was created by the orchestra—we should then have what is described as "perfect transmission". What we heard might be good or bad, pleasant or unpleasant, but it would at least be a faithful reproduction of what was played. Unhappily we live in an imperfect world in which perfect transmission is impossible.

Transmission is at its simplest when we sit in the same room as the orchestra. In this case the only transmitter is the air of the room, but even so the curve undergoes a good deal of distortion on its journey from the orchestra to our ears. For we shall see in a later chapter how the walls, the roof and floor, the clothes of the audience, and even the empty seats, all reflect sound in varying degrees, so that a considerable part of the sound we hear may have been reflected dozens of times before it reaches our ears, and every reflection will have changed the character of the sound-curve. If the music is broadcast, many more changes intervene before the sound reaches our ears. The sound-curve produced by the orchestra must then be handed on from the air of the concert room to the diaphragm of a microphone, from this to an electric current in a wire, from this to a shower of electrons jumping through a system of valves, from this to a current in an aerial, from this to waves of electric and magnetic force travelling through space, from these to another aerial and its connections in a receiving set, from these through more showers of electrons in valves to yet another current in another wire, from this to the diaphragm of a loudspeaker, from this to the air, and finally from this to our ear-drums. Each time the sound-curve is passed on from one of these carriers to another, it undergoes a certain amount of "distortion"; its shape is changed, its refinements and subtleties usually being more or less blurred over, imperfections and impurities creeping in, and the quality generally undergoing a change for the worse.

These changes are, however, insignificant in comparison with the changes introduced by the ear itself. We shall see later that this may add entirely new musical notes to those which are played by the orchestra. It may also— and this not only with people who are partially deaf— filter out certain other notes of high and low pitch entirely, refusing to transmit them to the brain. Even if it does not do this, it invariably favours certain sounds at the expense of others, so that the various sounds are heard in proportions quite different from those in which they were played by the orchestra.

We can now see the general plan of the discussion which lies before us. We have to consider the generation of sound, its passage to our ear-drums, and its transmission from these to our brains. We have seen that all sound, whether pleasant or unpleasant, whether music or mere noise, is represented by a curve. We shall first examine the general properties of such a curve, trying to discover why some curves produce pleasure when they reach our ears and some pain. We must then consider the transmission of sound, discussing how best to retain the pleasurable qualities in our sound-curve, as it passes on from one carrier to another, and how far it is possible to prevent unpleasant qualities contaminating the curve. Finally we shall have to discuss the strange transformations that the sound-curve may undergo inside our heads. In accordance with this, the next three chapters deal with various methods of producing sound, and the qualities of the sounds they produce, as indicated by their sound-curves. The next (Chapter v) will deal with the choice of the sound to be produced. After this we discuss in Chapter vi the transmission of sound from its source to the ear-drum, and in Chapter vii its transmission from the ear-drum to the brain.

CHAPTER II

TUNING-FORKS AND PURE TONES

We have seen that every sound, and every succession of sounds, can be represented by a curve, and our first problem must obviously be to find the relation between such a curve and the sound or sequence of sounds it represents—in brief, we must learn to interpret a sound-curve.

Pure Tones

Let us start by taking an ordinary tuning-fork as our source of sound. We begin with this rather than, let us say, a violin or an organ-pipe, because it gives a perfectly pure musical note, as we shall shortly see. If we strike its prongs on something hard, or draw a violin-bow across them, they are set into vibration. We can see that they are in vibration from their fuzzy outline. Or we can feel that they are in vibration by touching them with our fingers, when we shall experience a trembling or a buzzing sensation. Or, without trusting our senses at all, we may gently touch one prong with a light pith ball

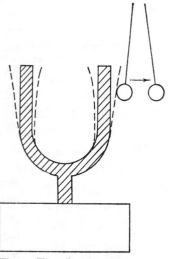

Fig. 5. The vibrations of a tuning-fork give a fuzzy appearance to the prongs and cause them to repel a light pith ball with some violence.

suspended from a thread, and shall find that the ball is knocked away with some violence.

When the prongs of the fork vibrate, they communicate their vibrations to the air surrounding them, and this in turn transmits the agitation to our ear-drums, with the result that we hear a sound. We can verify that the air is necessary to the hearing of the sound by standing the vibrating fork inside an air-pump and extracting the air. The fuzzy appearance of the prongs shews that the fork is

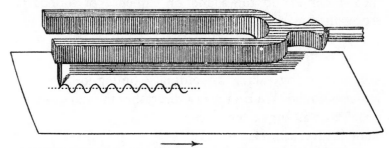

Fig. 6. The trace of a vibrating fork can be obtained by drawing a piece of paper or smoked glass under it.

still in vibration, but we can no longer hear the sound, because the air no longer provides a path by which the vibrations can travel to our ears.

To study these vibrations in detail, we may attach a stiff bristle or a light gramophone needle to the end of one prong of the fork, and while the fork is in vibration, run a piece of smoked glass under it as shewn in fig. 6, taking care that it moves in a perfectly straight line and at a perfectly steady speed. If the fork were not vibrating, the point of the needle would naturally cut a straight furrow through the smoky deposit on the glass; if we held the glass up to the light, it would look like fig. 7. In actual fact, we shall find it looks like fig. 8, which is a copy of an actual photograph; the vibrations have left their record in the smoke, so that the needle has not cut a straight but a

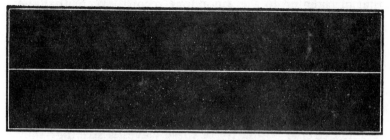

Fig. 7. The trace of a non-vibrating fork.

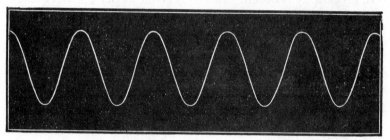

Fig. 8. The trace of a vibrating fork. The waves are produced by the vibrations of the fork, one complete wave by one complete vibration.

wavy furrow. Each complete wave obviously corresponds to a single to-and-fro motion of the needle point, and so to a complete vibration of the prong of the tuning-fork.

This wavy curve must clearly be the sound-curve of the sound emitted by the vibrating fork. For if we reverse the motion and compel the needle to follow the furrow, the sideways motions of the needle will set up similar motions in the prong to which it is attached, and these will produce exactly the same sound as was produced when the fork vibrated freely of itself. In fact, the whole process is like that of listening to a gramophone record, except that the tuning-fork, instead of a mica diaphragm, transmits the sound-vibrations to the air.

This simple experiment has disclosed the relation between the musical sound produced by a tuning-fork and its curve, which we now find to consist of a succession of similar waves.

The extreme regularity of these waves is striking; they are all of precisely the same shape, so that their lengths are all exactly the same, and they recur at perfectly regular intervals. Indeed, it is this regularity which distinguishes music from mere noise. So long as a gramophone needle is moving regularly to-and-fro in its groove we hear music; the moment it comes upon an accidental scratch on the record, so that its motion experiences a sudden irregular jerk, we hear mere noise. In such ways as this, we discover that regularity is the essential of a musical sound-curve. Yet the regularity can be overdone, and absolute unending regularity produces mere unpleasing monotony. The problem of designing a curve which shall give pleasure to the ear is not altogether unlike that of designing a building which shall give pleasure to the eye. A mere collection of random oddments thrown together anyhow is not satisfying; our aesthetic sense calls for a certain amount of regularity, rhythm and balance. Yet these qualities carried to excess produce monotony and lifelessness—the barracks in architecture and the dull flat hum of the tuning-fork in music.

Period, Frequency and Pitch

When a tuning-fork is first set into vibration, we hear a fairly loud note, but this gradually weakens in intensity as the vibrations transfer their energy to the surrounding air. Unless the fork was struck very violently in the first instance, we notice that the pitch of this note remains the same throughout; if the fork sounded middle C when it was first struck, it will continue to sound this same note until its sound dies away into silence.* On taking a trace of the whole motion, in the manner shewn in fig. 6, we find that the waves slowly decrease in height as the sound diminishes in strength, but they remain always of the same length.

If we measure the speed at which the fork is drawn over the smoked glass in taking this trace, we can easily calculate the amount of time the needle takes to make each wave. This is, of course, the time of a single vibration of the fork, and is only a minute fraction of a second; we call it the "period" of the vibration. The number of complete vibrations which occur in a second is called the "frequency" of the vibration. Actual experiment shews that a tuning-fork which is tuned to middle C of the pianoforte will be found to execute 261 vibrations in a second, regardless of whether the sound is loud or soft.

This frequency of 261 is associated with the pitch of middle C not only for the sound of a tuning-fork, but also for all musical sounds, no matter how they are produced. For instance, a siren which runs at such a rate that 261 blasts of air escape in a second will sound middle C. Or we may hold the edge of a card against a rotating toothed

---

* If the fork was struck very violently in the first instance, there may be a very slight sharpening of pitch as the vibrations become of more usual intensity.

wheel; if 261 teeth strike the card every second we again hear middle C. If a steam-saw runs at such a rate that 261 teeth cut into the wood every second, it is again middle C that we hear. The hum of a dynamo is middle C when the current alternates at the rate of 261 cycles a second, and this is true of all electric machinery. There are electric organs on the market in which the sound of a middle C pipe is copied, sometimes very faithfully, by an electric current which is made to alternate at the rate of 261 cycles a second. Again, when a motor-car is running at such a rate that the pistons make 261 strokes a second, a vibration of frequency 261 is set up, and we hear a note of pitch middle C in the noise of the engine.

All this shews that the pitch of a sound depends only on the frequency of the vibration by which it is produced. It does not depend on the nature of the vibration. Thus we may say that it is the frequency of vibration that determines the pitch of a sound. If there is no clearly defined frequency, there is no clearly defined pitch, because the sound is no longer musical.

When a siren or steam-saw or dynamo is increasing its speed, the sound we hear rises in pitch, and conversely. Thus we learn to associate high pitch with high frequency, and vice versa. If we experiment with a series of forks tuned to all the notes in the middle octave of the piano, we shall find the following frequencies:

c	261·6	f	349·2	a	440·0
c♯	277·2	f♯	370·0	a♯	466·2
d	293·6	g	392·0	b	493·9
d♯	311·1	g♯	415·3	c′	523·2
e	329·6				

Such, at least, are the numbers of vibrations for tuning-forks or any other instruments tuned in "equal temperament" (see p. 174, below) to the new (1939) internationally agreed pitch of a = 440. But many other standards of pitch are still in use, and even more were in use in the past. The lengths of old organ-pipes give us information as to the pitches which were in use in early times, and shew that one and the same note often had very different frequencies in different instruments. In Germany, for instance, Silbermann's great organ in Strassburg Cathedral (1713) had the pitch a = 393; while Schnitger's organ in S. Jacobi in Hamburg (1688) was tuned to a = 489—nearly four semitones higher. And this was not the worst, for the "church pitch" of Northern Germany had been given by Pretorius (1619) as a = 567, which is a full six semitones higher than that of the Strassburg organ. Not only so, but secular music was often played in a substantially higher pitch than sacred music, so that there was a "chamber" pitch which was quite distinct from the "church" pitch. There were similar variations in other countries. In England, Father Smith's organ in Trinity College, Cambridge, was tuned (1759) to a = 395, while Berhardt Schmidt's organs in Durham Cathedral and the Chapel Royal gave a = 474·1—more than three semitones above the Trinity organ.

Early in the eighteenth century efforts were made to make musical pitch more uniform, but it still ranged from about 415 to 430 for a. It stayed fairly stationary within these limits for about a century, when musicians, striving for greater brilliance and keenness of tone, began again to raise the frequency. In 1879 the Covent Garden Orchestra was playing at a pitch of a = 450, while in America the so-called "Concert Pitch" went as high as a = 461·6.

In 1859 a French Government Commission recommended a = 435 as standard pitch, and this came into fairly general use on the Continent. But in America, a = 440 was often taken as standard, and in England, where it has been usual to tune from c rather than a, c′ = 522, corresponding to a = 438·9. In 1939, an international conference met in London and agreed on a = 440 as a new standard for universal use, at least in broadcasting. With this standard the frequencies of notes are those given in the table on p. 22.

These frequencies might at first sight be thought to be a mere random collection of numbers, but a little study shews that they are not.

We notice at once that the first number 261·6 is just half of the last number 523·2. Thus our experiments have shewn that in this particular case the interval of an octave corresponds to a 2 to 1 ratio of frequencies, and other experiments shew that this is universally true—doubling the frequency invariably raises the pitch by an octave. The octave interval is fundamental in the music of all ages and of all countries; we now see its physical significance.

We may further notice that the interval from c to c♯ represents a rise in frequency of just about 6 per cent., and a little arithmetic will shew that the same is true for every other interval of a semitone. The rise cannot be precisely 6 per cent. for each semitone, since if it were, the rise in the whole octave, consisting of twelve such intervals, would be equal to 1·06 × 1·06 × 1·06 × ...etc., there being twelve factors in all, each equal to 1·06. This is the quantity which the mathematician describes as $(1 \cdot 06)^{12}$, and it is equal to $2 \cdot 0122$, and not to exactly 2.

In an instrument such as the piano or organ, which is tuned to "equal temperament" (see p. 174, below), the exact interval of 2 is spread equally over the twelve semi-tone intervals which make the octave. Each step accordingly represents a frequency ratio of $1 \cdot 05946$, since this is the exact twelfth root of 2.

By repeated multiplication by the factor $1 \cdot 05946$, we obtain the following table for the ratios of the frequencies of notes of any octave to that of c:

Frequency ratios within the octave

$$c = 1$$

$c\sharp = 1 \cdot 05946$	$g = (1 \cdot 05946)^7 = 1 \cdot 4983$
$d = (1 \cdot 05946)^2 = 1 \cdot 1225$	$g\sharp = (1 \cdot 05946)^8 = 1 \cdot 5874$
$d\sharp = (1 \cdot 05946)^3 = 1 \cdot 1892$	$a = (1 \cdot 05946)^9 = 1 \cdot 6818$
$e = (1 \cdot 05946)^4 = 1 \cdot 2599$	$a\sharp = (1 \cdot 05946)^{10} = 1 \cdot 7818$
$f = (1 \cdot 05946)^5 = 1 \cdot 3348$	$b = (1 \cdot 05946)^{11} = 1 \cdot 8877$
$f\sharp = (1 \cdot 05946)^6 = 1 \cdot 4142$	$c' = (1 \cdot 05946)^{12} = 2 \cdot 0000$

The next note $c\sharp'$ will, of course, have a frequency of $(1 \cdot 05946)^{13}$ times that of c. Since $(1 \cdot 05946)^{12} = 2$, this is the same thing as $2 \times 1 \cdot 05946$ or twice the frequency of $c\sharp$, and so on.

Increasing the frequency of any note whatever by a factor $1 \cdot 05946$ simply raises its pitch by a semitone, and this is true throughout the whole of the scale. We can verify this by increasing the speed of a siren or a steam-saw or any of the other appliances already mentioned. Perhaps the simplest way of all is to take a gramophone record of a pianoforte solo which has been recorded for the standard rate of 78 revolutions, and run it at 82·6 revolutions, which is just 1·05946 times the standard rate. We shall find that the whole piece sounds exactly as it did when we played it at the standard rate of 78, except that it is a semitone higher. If the original piece was in the key of C we now hear it in the key of C♯, and so on. If we could run our gramophone at double its normal rate, 156 revolutions, we should again hear our piece of music in its original key of C, but played an octave higher.

We can now make a table shewing the frequency of every musical note. Before doing this, it will be well to introduce a notation to distinguish the different octaves, and we shall find it convenient, whenever we wish to specify particular octaves, to use a slight modification of a notation originally proposed by Helmholtz. In this, the different octaves are distinguished by their notes being printed in different styles of type, each octave being supposed to start with c, and to extend to the b above.

The styles of type used in the present book are shewn in fig. 9 opposite.

It will be noticed that the unaccented letters c, d, e,... are not used, so that we are free to use them when it is not desired to specify any particular octave.

We must also decide on our standard of pitch. Partly because there are so many different pitches in use, and partly for numerical convenience, it is usual to make all theoretical calculations in terms of a standard pitch $c'' = 512$. If we use this pitch, our table stands as below. The frequencies for any pitch in actual use can be obtained by adding a small percentage. For $c'' = 522$, for instance, we must add 2 per cent., since 522 is very nearly 2 per cent. greater than 512.

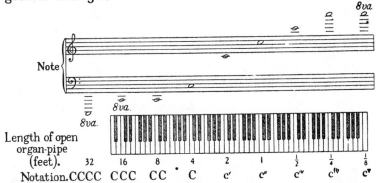

Fig. 9. The notation used in the present book, which is a slight modification of that of Helmholtz, shewn in relation to the musical staves, the keyboard of the pianoforte, and the lengths of open organ-pipes.

Frequencies of tones from CCCC *to* c^{vi} (c″ = 512)

Octave Note	CCCC	CCC	CC	C	c′	c″	c‴	c^{iv}	c^{v}
c	16	32	64	128	256	512	1024	2048	4096
c♯	17	34	68	136	271	542	1085	2170	4340
d	18	36	72	144	287	575	1149	2299	4598
d♯	19	38	76	152	304	609	1218	2436	4871
e	20	40	81	161	323	645	1290	2580	5161
f	21	43	85	171	342	683	1367	2734	5468
f♯	23	45	91	181	362	724	1448	2896	5793
g	24	48	96	192	384	767	1534	3069	6137
g♯	25	51	102	203	406	813	1625	3251	6502
a	27	54	108	215	431	861	1722	3444	6889
a♯	29	57	114	228	456	912	1825	3649	7298
b	30	60	121	242	483	967	1933	3866	7732

Simple Harmonic Curves

Having learned all we can from the regularity and length of the waves in fig. 8, let us next examine their form. The extreme simplicity of their shape is very noticeable, although it must be said at once that this is not a property of all sound-curves; these particular curves are simple because they are produced by the simplest of all musical instruments—the tuning-fork. Exact measurement shews that the curve has a shape with which the mathematician is very well acquainted. It is called a "sine" curve, or a "simple harmonic" curve, while the motion of the needle which produces it is described as "simple harmonic motion".

These simple harmonic curves and the simple harmonic motion by which they are produced are of fundamental importance in all departments of mechanics and physics, as well as in many other branches of science. They are particularly important in the theory of vibrations, and this makes them of especial interest in the study of music, since musical sound is almost invariably produced by the vibrations of some mechanical structure—a stretched string, a column of air, a drum-skin, or some metallic object such as a cymbal, triangle, tube or bell. For this reason, we shall discuss vibrations in some detail.

General Theory of Vibrations

Generally speaking, every material structure can find at least one position in which it can remain at rest— otherwise it would be a perpetual motion machine. Such a position is called a "position of equilibrium". When a structure is in such a position, the forces on each particle of it—as for instance the weight of the particle, and the pushes and pulls from neighbouring particles—are exactly balanced. Any slight disturbance, such as a push, pull or knock from outside, will cause the structure to move out of this position of equilibrium to some new position, in which the forces on a particle are no longer evenly balanced; each particle then experiences a "restoring force" which tends to pull it back to the position it originally occupied.

This force starts by dragging the particle back towards its original position of equilibrium. In time it regains this position, but as it is now moving with a certain amount of speed, it overshoots the position and travels a certain distance on the other side before coming to rest. Here it experiences a new force tending to pull it back; again it yields to this force, gets up speed, overshoots the mark, and so on, the motion repeating itself time after time. Clearly the trace of the motion of any particle will be a succession of waves, like those we have already obtained from the tuning-fork in fig. 8 (p. 19).

Motion of this kind is described by the general term "oscillation". In the special case in which each particle only moves through a very small distance, the motion is called a "vibration". Thus a vibration is a special kind of oscillation, and, as it happens, possesses certain very simple properties which are not possessed by oscillations in general. It is usually true of oscillations that the farther a particle moves from its position of equilibrium, the greater is the restoring force pulling it back. But in a vibration the restoring force is *exactly proportional* to the distance the particle has moved from its position of equilibrium; draw it twice as far from this position, and we double the force pulling it back.

A simple mathematical investigation shews that when this relation holds, the motion of every particle will be of the same kind, whatever the structure to which it belongs.

Motion of this kind is defined to be "simple harmonic motion".

We have already found a concrete instance of this kind of motion in the tuning-fork. Another is provided by what is perhaps the simplest mechanical structure we can imagine —a weight suspended by a fine thread. The position of equilibrium is one in which the weight lies at a point C exactly under the point of suspension. When the weight is drawn a short distance aside to an adjacent position B, there is no longer equilibrium, and the weight tends to fall back to C. In technical language, a restoring force acts on the weight, tending to draw it back to its position of equilibrium C, and it is a simple problem in dynamics to find its amount. So long as the displacement of the weight is not too large, we find that the restoring force is exactly proportional to the extent of the displacement BC, so that the condition for simple harmonic motion is fulfilled. Indeed, if we take a trace by attaching a needle to the weight and running a piece of paper horizontally under it, as in fig. 11, we shall find that this trace is a simple harmonic curve exactly like that made by our tuning-fork.

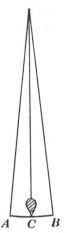

Fig. 10. A position of equilibrium. The weight can rest in equilibrium at C but nowhere else. If we pull it aside to B, it tends to return to C.

If we set our suspended weight swinging more violently, and again take a trace of its motion, we shall again obtain a simple harmonic curve. The waves will, of course, be greater in size, but their period will be exactly the same as before. We find that the swinging weight makes just as many swings per second, no matter what the extent of these swings may be, provided always that they are small enough to qualify as vibrations. This illustrates the well-known fact that the period of vibration of a pendulum depends only

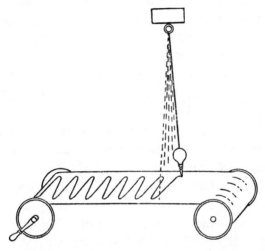

Fig. 11. Taking the trace of a swinging pendulum. The trace is found to be a simple harmonic curve, exactly similar to that given by a vibrating tuning-fork (fig. 6).

on its length, and not on the extent of its swing; it is because of this that our pendulum clocks keep time.

We found a similar property in the tuning-fork, the period of its vibrations being the same whether we struck it fairly hard or only very softly. And all true vibrations possess the same property—the period is independent of the extent and energy of the swing. This is a most important fact for the musician. It means that every musical instrument in which the sound is produced by vibrations will "keep time" like a pendulum clock, and so will give a note of the same frequency, and therefore of the same pitch, whether it is played soft or loud. Without this property it may almost be said that music, as we know it, would be impossible. We can hardly imagine an orchestra acquitting itself with credit if every note was out of tune unless it was played with exactly the right degree of force. Crescendos and diminuendos could only be produced by adding and subtracting instruments. As the note of a piano or any percussion instrument decreased in strength it would also change in pitch, and every piece would inevitably begin with a howl and end with a wail.

At the same time, every musician is familiar with cases in which the pitch of an instrument is changed appreciably by playing it softer or louder. The flautist can always pull his instrument a bit out of tune by blowing strong or weak, while the organist knows only too well the dismal wail of flattened notes which is heard when his wind gives out. We shall discuss the theory of such sounds as these later, and shall find that they are not produced by absolutely simple vibrations like those of the tuning-fork or pendulum.

Simultaneous Vibrations

Many structures are capable of vibrating in more than one way, and so may often be performing several different vibrations at the same time. There is a very general principle in mechanics, which asserts that when any structure whatever is set into vibration—provided only that the displacement of each particle is small—the motion of every particle is either a simple harmonic motion or else is a more complicated motion which results from superposing a number of simple harmonic motions, one for each vibration which is in progress.

A simple illustration will shew how this can be. Let us suppose that while our tuning-fork is in vibration we hit it on the top of one of the prongs with a hammer. We shall hear a sharp metallic click, which is known as the "clang

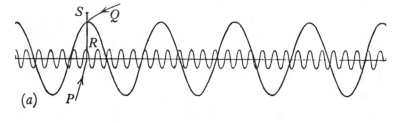

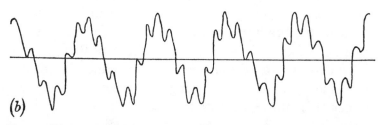

Fig. 12. The superposition of two vibrations. The two wavy curves in (a) have periods which stand in the ratio of 6¼ to 1. On superposing them we obtain the curve (b), which represents very closely the sound-curve of a tuning-fork which is sounding its clang tone.

tone" of the fork. A good musical ear may perhaps recognise that its pitch lies about 2½ octaves above the ordinary note of the fork. Clearly the blow of the hammer has started new vibrations in the fork, of much higher frequency than the original vibration. If we had taken a trace of the motion when the original vibration was acting alone we should have obtained a curve like that shewn in fig. 1 of Plate II. This is reproduced as the long-waved curve in fig. 12 (a). If we take a trace of the clang tone alone, it will be like the short-waved curve in fig. 12 (a), this representing a simple harmonic motion having $6\frac{1}{4}$ times the frequency of the main vibration.

Now suppose we take a trace when the two vibrations are going on together. At the instant of time represented at the point P, the particle under consideration is displaced through a distance PQ by the main vibration, and through a distance PR by the vibration which produces the clang tone. Thus the operation of the two vibrations together displaces it through a distance $PQ+PR$, and this is equal to PS if we make QS equal to PR. By adding together displacements in this way all along the curve, we obtain the curve shewn in fig. 12 (b) as the trace to be expected when both vibrations are in action together. The photograph of an actual trace is shewn in fig. 2 of Plate II.

In addition to the clang tone just mentioned, we may often hear a second clang tone about four octaves higher than the fundamental note of the fork. Indeed, it is difficult to start the fork sounding in such a way that the pure tone of the fork is heard without any admixture of these higher tones. We more usually obtain a mixture of all three tones, but this does not interfere with the utility of the tuning-fork as a source of pure musical tone, since the sounds of higher frequency die away quite rapidly, and the ear soon hears nothing but the fundamental tone of the fork.

A second example of simultaneous vibrations can be made to teach us something new. If we return to our weight suspended by a string and knock it sideways, it

PLATE II

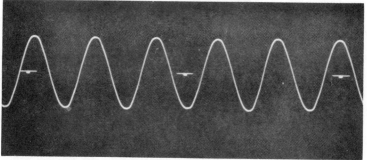

Dayton C. Miller

Fig. 1. The sound-curve of the simple tone from a tuning-fork. The note is of frequency 256 (middle C), and the dots indicate intervals of $\frac{1}{100}$ second.

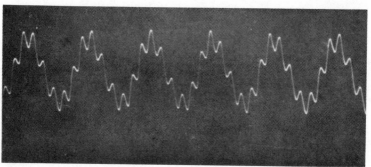

Dayton C. Miller

Fig. 2. The sound-curve of the note from a tuning-fork when the clang tone is sounding. The clang tone superposes small waves onto the longer waves, shewn in fig. 1 above, which represent the main tone of the fork.

SOUND-CURVES OF A TUNING-FORK

will swing from side to side pendulum-wise through some such path as AB in fig. 10 (p. 30), and its motion, as we have already seen, will be simple harmonic motion. Suppose, however, that when the weight is at B, we give it another slight knock in the direction at right angles to AB, i.e. *through* the paper of our page in fig. 10. This sets up a new vibration in a direction at right angles to AB, and the motion in this direction also must be simple harmonic motion. As we have seen that the period of a pendulum

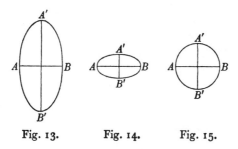

Fig. 13.   Fig. 14.   Fig. 15.

Figs. 13, 14 and 15. Three different types of motion which can be executed by the bob of a conical pendulum.

depends only on its length, the new motion will have the same period as the original motion. The whole motion is accordingly obtained from the superposition of two simple harmonic motions whose periods are equal.

If we watch the weight from a point directly above it, we shall see it moving in a curved path round its central position C. If the second knock was violent, its path will be an elongated ellipse such as $AA'BB'$ in fig. 13. If the knock was gentle, its path will be an ellipse elongated in the other direction such as $AA'BB'$ in fig. 14. But if the knock was of precisely the same strength as that which originally set the pendulum in motion along AB, then the weight will move in the circle $AA'BB'$ in fig. 15, forming the arrangement which is generally described as a conical pendulum. It must move with the same speed at each point of its journey, for it is moving in a perfectly level path, so that there is no reason why it should move faster at any one point than at any other.

Thus we learn that each of the motions illustrated in figs. 13, 14 and 15 can be regarded as the superposition of two simple harmonic motions of equal periods. The last of the three is by far the most interesting, because it shews us that a simple circular motion performed at uniform speed can be regarded as made up of two simple harmonic motions in directions at right angles to one another. To put this more definitely, let us imagine that the point P in fig. 16 moves round the circle $AA'BB'$ with uniform speed, like the hand of a clock. Wherever P is, let us draw perpendiculars PN, PM on to the lines AB, $A'B'$. Then, as P moves steadily round the circle, N moves backwards and forwards along AB, while M moves backwards and forwards along $A'B'$. We have learnt that the motion of each of these points will be simple harmonic motion.

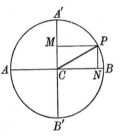

Fig. 16. A geometrical interpretation of simple harmonic motion. As the point P moves steadily round the circle, the point N moves backwards and forwards along AB, and its motion is simple harmonic motion.

This gives us a simple geometrical explanation of simple harmonic motion—as P moves steadily round in a circle, the point N moves in simple harmonic motion. It is easy to see from this definition that the motion of the piston in the cylinder of a locomotive or a motor-car must be approximately simple harmonic motion.

Or we may look at the problem from the other end, and see that as the point N moves to-and-fro in simple harmonic motion along AB, the point P moves steadily round the circle $AA'BB'$. This circle is called the "circle of reference" of the simple harmonic motion. Its diameter AB is called the "extent" of the motion, while its radius CA or CB is called the "amplitude" of the motion.

Energy

The amplitude of a vibration gives an indication of its energy, for it is a general law that the energy of a vibration is proportional to the square of the amplitude. For instance, a vibration which has twice the amplitude of another has four times the energy of the other; in other words, the vibrating structure to which it belongs has four times as much capacity for doing work stored up within itself, and it must get rid of this in some way or other before it can come to rest. The energy stored up in a musical instrument is usually expended in setting the air around it into vibration; indeed it is only through its steady outpouring of energy into the surrounding air that we hear the instrument at all.

It follows that if we want to maintain a vibration at the same level of energy we must continually supply energy to it—as we do with an organ-pipe or a violin-string. If energy is not supplied the vibration will die away—as with a piano-string or a bell or a cymbal. The amplitude of the vibration then slowly decreases, and the circle of reference shrinks in size.

When a structure is performing several vibrations at the same time, energy does not usually pass from one vibration to another. The vibrations are independent, each possessing its own private store of energy which it preserves intact, except for what it may pass on to other outside structures—as for instance, the air around it. Thus the energy of a number of simultaneous vibrations may be thought of as the sum of the energies of the separate vibrations.

Simultaneous Sounds

When a tuning-fork is sounding, every particle of its substance moves in simple harmonic motion, and those particles which form its surface transmit their motion to the surrounding air. The final result is that every particle of air which is at all near to the tuning-fork is set into motion and moves with a simple harmonic motion, which will naturally have the same period as the tuning-fork. This period is still preserved when the vibration is passed on to the ear-drum of a listener—that is why the note heard by the ear has the same pitch as the fork.

A more complicated situation arises when two tuning-forks are standing side by side. Each then imposes a simple harmonic motion on to the particle of air, so that this has a motion which is obtained by superposing the two motions.

We must study motions of this kind in some detail, because they are of great importance in the practical problems of music. We begin with the simplest problem of all—the superposition of two motions which have the same period. The resulting motion is that which would be forced on a particle of air by the simultaneous vibrations of two forks of the same pitch standing side by side.

Superposing Vibrations of the same Period

The two simple harmonic motions can be represented by two simple harmonic curves, such as those which pass through X and Y in fig. 17. These particular curves have been drawn with their amplitudes in the ratio of 5 to 2, so that $YN = \frac{2}{5} XN$, and the same relation holds all along the

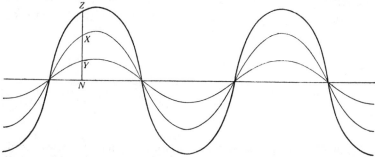

Fig. 17. The superposition of two simple harmonic motions of equal period. Here the vibratory motions (represented by the thin curves) are "in the same phase"—crest over crest and trough over trough. The vibrations now reinforce one another, and their resultant (represented by the thick curve) has an amplitude which is equal to the sum of the amplitudes of the two constituents.

curves. At the instant of time represented at the point N, the first harmonic motion produces a displacement through a distance XN, while the second produces a displacement through a distance YN which is $\frac{2}{5}$ times XN. Thus the combined effect of the two motions is a displacement through a distance equal to $1\frac{2}{5}$ times XN. This is represented by ZN in fig. 17.

The thick curve through Z is drawn so that its distance above or below the central line is everywhere exactly $1\frac{2}{5}$ times that of the thin curve through X. This curve must then represent the motion of which we are in search. It is simply the thin curve through X magnified $1\frac{2}{8}$ times vertically, while its horizontal dimensions remain unchanged. Thus the new motion is a simple harmonic motion having an amplitude equal to the sum of the amplitudes of the constituent motions, and the same period as both.

The foregoing instance is only a very special case of the general problem, for the thin curves in fig. 17 are drawn in a very special way. The crests of the waves of the two curves

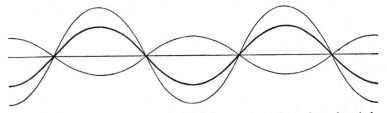

Fig. 18. The superposition of two simple harmonic motions of equal period. Here the vibratory motions (represented by the thin curves) are "in opposite phase"—crest over trough and trough over crest. The constituent vibrations now pull in opposite directions, and so partially neutralise one another, the amplitude of their resultant (represented by the thick curve) being equal to the difference of the amplitudes of the two constituents.

occur at the same instants, as also the troughs; in the diagram, crest lies directly over crest and trough over trough. Vibrations in which this relation holds are said to be "in the same phase".

The curves might equally well have been drawn as in fig. 18, the crests of one set of waves occurring at the same instants as the troughs of the other set. Vibrations in which this relation holds are said to be "in opposite phase". Crest lies over trough and vice versa, so that the two constituents produce displacements in opposite directions. The resultant motion is again that shewn in the thick curve, but its amplitude is no longer $(1 + \frac{2}{5})$ times the amplitude of the larger constituent, but only $(1 - \frac{2}{5})$ times.

We must not, however, expect as a matter of course that two motions which occur simultaneously will be either in the same, or in opposite, phase. Such simplicity is unusual, and it is far more likely that the crests of one set of waves will be neither over the crests nor over the troughs of the other set, but somewhere in between, as shewn in fig. 19. If we add together the displacements represented

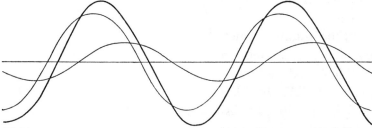

Fig. 19. The superposition of two simple harmonic motions of equal period. Here there is no simple phase relation between the two constituent vibratory motions (represented by the thin curves), but their resultant is still a simple harmonic motion (represented by the thick curve).

by the two thin curves here, using the method illustrated in fig. 17 (i.e. making $ZN = XN + YN$, and so on), we shall find that the resultant motion is represented by the thick curve shewn in the figure. We may judge by eye that this is yet another simple harmonic curve, as in actual fact it is, but we can only prove this by a new method of attack on the problem, to which we now turn.

We have seen that any simple harmonic motion can be derived from the steady motion of a point round a circle. For instance, as the point P moves round the circle in fig. 16, the point N moves backwards and forwards along

the line *AB* in simple harmonic motion. The two simple
harmonic motions which we now want to superpose can of
course be derived from the motions of two points, each
moving steadily round a circle of its own. Let the two
points be *P* and *Q* in
fig. 20, so that the points
N, *O* immediately be-
neath them execute the
simple harmonic motions
with which we are con-
cerned.

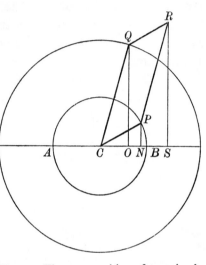

At the instant to which
fig. 20 refers, the motion
of *P* has produced a displacement *CN*, while that
of *Q* has produced a displacement *CO*, so that
the total displacement,
being the sum of the two,
is equal to *CO + CN*.

To represent this in
fig. 20, we start from *Q*,
and draw the line *QR* in

Fig. 20. The superposition of two simple
harmonic motions. As *P* and *Q* move round
their respective circles, *N* and *O* execute
simple harmonic motions. The resultant
motion is that executed by *S*, because
CO + CN = CS.

a direction parallel to *CP* and of length equal to *CP*.
Then, because *QR* and *CP* are parallel and equal, the
length *OS* which lies directly under *QR* must be exactly
equal to the length *CN* which lies directly under *CP*.
Hence the sum we need, namely *CO + CN*, must be equal
to *CO + OS*, and so to *CS*.

Thus as *P* and *Q* move round their respective circles,
the points *N* and *O* execute the two constituent simple

harmonic motions, and the point S executes the motion which results from their superposition.

We are at present supposing the two simple harmonic motions performed by N and O to be of the same frequency, so that the radii CP and CQ rotate at exactly the same rate and the angle PCQ remains always the same. Indeed, we can visualise the whole motion by imagining that we cut the parallelogram $CPRQ$ out of cardboard, and then make it rotate round C at the same rate as P and Q. We see that R will move in a circle at uniform speed, so that S will move backwards and forwards along AB in simple harmonic motion. This shews that when two simple harmonic motions have the same frequency, the result of superposing them is a third simple harmonic motion of the same frequency as both. In terms of music, the simultaneous sounding of two pure tones of the same pitch produces a pure tone which is still of the same pitch.

Loudness

The loudness of this sound is a matter of some interest. The resultant vibration has an amplitude which is represented by the length of the line CR, and this depends not only on the amplitudes CP, CQ of the constituent vibrations, but also on the angle between CP and CQ, being large when the angle is small, and vice versa. As we have already seen, the energy of the sound is proportional to the square of CR, so that the sound will be loudest when CR is as large as possible, and this occurs when CP and CQ lie in the same direction. The parallelogram $CPRQ$ then collapses into a straight line, and the amplitude CR of the new vibration is simply the sum of the amplitudes CP

and CQ of the constituent vibrations. This is the case, already illustrated in fig. 17, in which the vibrations are in the same phase; they reinforce one another continually, and to the fullest possible extent.

The sound will be least loud when CR has its smallest possible value. This occurs when CP and CQ lie in opposite directions, the amplitude CR then being the difference of the amplitudes CP and CQ of the two constituents. This is the case, already illustrated in fig. 18, in which the vibrations are in opposite phase; as they continually pull in exactly opposite directions, they neutralise one another in the highest degree possible.

The angle PCQ is equal to zero when the vibrations are in the same phase, and is equal to 180°, or two right angles, when they are in opposite phase. More usually, the angle PCQ will have some intermediate value, and we say that the vibrations have a "phase difference" PCQ, or more precisely that the phase of Q is PCQ behind that of P. This case has been illustrated graphically in fig. 19.

Interference of Sound

Our first result has shewn that if we could arrange for two notes to be sounded in exactly equal strength and in exactly the same phase, their resultant would have double the amplitude of either, and so would have four times the energy of either of its constituents. This means that each note would give out double as much energy in combination as when sounded alone, and it might seem at first sight that power could be gained by subdividing our vibrating mechanism into smaller units—using, for instance, two weak organ-pipes of the same pitch instead of one powerful pipe of double the strength of either. But this is not so—nature never gives us something for nothing, although she may seem to give us nothing for something.

In actual fact, if two similar organ-pipes are put side by side on the same wind-chest and blown together, the sound we hear will not have four times the energy of that produced by a single pipe; indeed if the mouths of the two pipes face one another, we shall hear little more than a sound of rushing air. Yet if we place a feather near to the lip of either pipe, it will flutter as strongly as if the pipe were being blown alone and producing its usual musical note. If we put one end of a rubber tube near to the lip of either pipe, and the other end in our ear, we shall find that the pipe is in actual fact emitting its usual musical note.

The explanation of the apparent paradox is as follows. As soon as the air in the first pipe starts its vibrations, the outflow of air from the mouth of the pipe creates an excess of pressure which tends to drive air into the mouth of the other pipe, or vice versa. Thus the pipes tend to get into such a condition that the air which streams out of one tends to stream into the other. In this way their vibrations get into opposite phases at the very outset, and when the vibrations are in opposite phases, their resultant is a new vibration of an amplitude which is equal, not to the sum of the amplitudes of the separate vibrations, but to their difference, i.e. nil. The pipes are accordingly said to "destroy one another's speech". In practice, the organ-builder will put two similar pipes as far apart as is conveniently possible, so as to reduce their mutual interference to a minimum.

The two prongs of a tuning-fork behave in somewhat the

same way. They may be regarded as two separate vibrators, and it is easily seen that their vibrations must necessarily tend to get into opposite phases, and so neutralise one another. Because of this, a tuning-fork can be made to sound louder by holding a card in such a position that it prevents the air vibrations streaming off one prong of the fork on to the other.

At first sight it seems very paradoxical that two sounds can cancel one another in the way just explained. We are apt to think of a sound as something which produces a sensation, and it then seems reasonable that if one sound produces a certain sensation, two sounds must produce twice as much sensation. And this would, of course, be true if the sounds ever got so far as producing sensation. In actual fact they cancel one another before getting any-where near to our ears—instead of each organ-pipe pro-ducing waves of sound which travel through the air to our ears, it merely produces waves which are sucked in by the other pipe, and vice versa, so that the auditory nerves need experience no sensation at all. Similar considerations apply to the case of two strings lying side by side and stretched so as to sound the same note, and explain why three strings to each note of the piano are better than two.

Beats

Let us next examine the result of sounding together two tuning-forks which are nearly, but not quite, in tune, so that the two simple harmonic vibrations which are to be superposed are nearly, but not quite, of the same frequency. We can represent this by imagining P and Q in fig. 20 (p. 42) to move round their respective circles at slightly different rates. In order to have a definite problem before us, let us suppose that P makes 520 complete revolutions a second, while Q makes 522. Then Q makes two complete revolutions more than P every second, and so is continually gaining on P, just as the minute-hand of a clock continually gains on the hour-hand. Because of this the angle PCQ is continually changing, and the parallelogram $CPRQ$ will no longer rotate about C like a rigid structure. In the position shewn in fig. 20, it is closing up, so that soon CP and CQ will coincide in direction. When this occurs, the two simple harmonic vibrations will be in the same phase, and the amplitude of the resultant vibration will be $CQ+CP$. After this the angle opens out again until, a quarter of a second later, CP and CQ point in exactly opposite directions. The two simple harmonic vibrations are now in opposite phase, so that the amplitude of the resultant is $CQ-CP$. Yet another quarter second later, and the constituent vibrations are again in the same phase, with a resultant once more of amplitude $CQ+CP$. Thus the resultant is a simple harmonic vibration whose amplitude continually varies between the limits $CQ+CP$ and $CQ-CP$.

If the amplitudes CP and CQ are exactly equal, the amplitude of the resultant disappears completely at the instants when the constituents are in opposite phase, so that the sound we hear consists of pulses of sound which occur regularly every half second, interleaved with moments of complete silence.

If the amplitudes CP and CQ are not exactly equal, the amplitude of the resultant sound continually fluctuates. Some sound can always be heard, but its loudness rises and falls, and this endows it with a wavy quality. The moments of maximum sound, or sometimes the whole intervals from minimum to minimum of the sound, are described as "beats".

In the particular experiment just described, the number of beats a second will be 2, because $522 - 520 = 2$. If the two notes had been twice as far out of tune—say of frequencies 524 and 520—we should have had four beats a second, and so on. The more out of tune the notes are, the more frequent the beats. When a tuner is bringing two strings of a piano or two pipes of an organ to the same pitch, he must tune until the beats can no longer be heard at all. Under the best conditions beats as slow as one in 30 seconds can be heard.

Cases such as those just mentioned, in which there are only two or four beats to the second, do not usually produce an unpleasant sound. Indeed certain registers of the organ, such as the "voix celeste" and the "unda maris", produce the effect deliberately by the device of each note having two pipes, which are purposely put sufficiently out of tune with one another to give two or three beats a second. The voix celeste is usually constructed of string-toned pipes. Its fantastic name notwithstanding, it attempts to represent the slightly undulating sound heard when the strings of an orchestra play in unison; the undulations arise in part from the "beats" which must necessarily occur since the instruments can never be in perfect tune with one another, but in still greater part from a more subtle cause which we shall discover when we study violin tone in detail (p. 102, below). The still more fantastically named unda maris usually consists of flute-toned pipes, and bears some resemblance to voices singing in attempted but imperfect unison. These undulations of sound endow organ tone with a certain quality of life and motion which is otherwise wanting.

Most ears find the sound of these stops agreeable, at any rate in the upper half of the keyboard; the stops seldom go below tenor C because the beats become less pleasant in the bass octave. Indeed it is a general rule that beats sound unpleasant when the number of beats per second is comparable with the frequency of the main tone. For instance, a 16-foot organ-pipe CCC gives 33 vibrations a second, while the adjacent CCC♯ pipe gives 35 vibrations a second. Thus if a 16-foot pedal-stop is drawn and the two bottom keys are depressed simultaneously, we hear two beats a second, and find the effect very unpleasant to the ear. If CCC and DDD are sounded together, we hear four beats a second, with still more unpleasant results.

To explore this effect more thoroughly, let us take a tuning-fork of frequency 261 (middle C of the piano) and sound with it in succession a series of forks of gradually ascending frequencies, say 262, 264, 266, 268, 270, and so on. With the 262 fork we hear one beat a second, and the effect is not unmusical, although the beats may be a bit irritating by their slowness. The next fork, of frequency 264, will give three beats a second, and the effect is still not unpleasing. The next fork, 266, with five beats a second, produces a distinctly less pleasant result, sounding more hurried and less musical, and this unpleasantness continually increases, until we reach the fork 284 which gives 23 beats a second. We can still recognise the beats as such, but their effect has become highly unpleasant—a confused

jangle of hurried sound rather than a musical note. After
this the beats remain audible for a time, but their unpleasantness diminishes. By the time we reach the fork 320
or thereabouts, the individual beats can no longer be
distinguished, but the sound is still unpleasant. The beats
do not return again, no matter how high we push the
pitch of the second fork. We shall discuss the meaning of
all this later (p. 153); in the meantime the following table
gives the result of experiments made by Mayer and Stumpf
with the object of discovering the greatest number of beats
which can be heard with pure sounds of different pitch.
In these experiments fork I is made to sound continuously,
while fork II is continually raised in pitch. The experi-
menter notices (a) the number of beats at which the
sensation is most unpleasant and (b) the greatest number
of beats which can be heard—i.e. the point beyond which
beats can no longer be heard.

Frequency of fork I	No. of beats per second at which		Interval in semi-tones until beats disappear
	beats are most unpleasant	beats can no longer be heard	
96	16	41	6
256	23	58	4
575	43	107	3
1707	84	210	2
2800	106	265	1·5
4000	—	400*	1·6

* This is the number given by Stumpf; Helmholtz had previously found a
limit of only 276 beats, or 1·1 semitones.

The unpleasantness seems to arise in part from the
mental irritation of trying to follow a succession of abrupt
and rapidly repeated changes, and in part from the purely

physical irritation produced by a succession of rapidly alternating stimuli. It is rather like the irritation we experience when watching a flickery cinematograph film. If the pictures follow one another at the rate of one per second, we feel no irritation, because the eye has ample time to adjust itself to each picture before this gives way to its successor. But this is not so when there are ten pictures to the second; we then get eye-strain and headache from the effort of trying to follow. If there are as many as twenty pictures to the second, the sequence merges into a continuous stream, and irritation gives place to satisfaction.

Difference and Summation Tones

Finally, let us examine the result of superposing two pure tones which are entirely out of tune, i.e. two simple harmonic vibrations of widely different frequencies. To make the problem definite, let us suppose that their frequencies are 600 and 800. Then we may suppose that in fig. 20 the directions of CP and CQ coincide at the start, and that P and Q rotate round their circles of reference at the rate of 600 and 800 revolutions a second respectively, so that Q passes and overtakes P 200 times every second. Each time that this happens, conditions are precisely the same as when the motion started, so that the motion repeats itself 200 times every second, and so shews a regular periodicity of frequency 200. Had this frequency been sufficiently low, the periodicity would have been audible in the form of beats. Although it is too high to be heard in this way, the periodicity is still latent in the mathematics of the problem, and, under conditions which we shall explore later, it may become audible as a "difference tone"—a tone whose frequency is the difference of the frequencies of the two tones which are superposed (p. 234).

From the point of view of obtaining a faithful representation of simple harmonic motion, it is a matter of complete indifference whether the moving point moves round the circle of reference in one direction or the other. Thus we can equally well represent the two motions which are to be superposed by supposing that in fig. 20 P and Q move round their circles in opposite directions. With the frequencies we have taken, they now pass one another 1400 times a second. Thus we see that the motion which results from superposing the two frequencies of 600 and 800 has latent in it yet another periodicity of frequency 1400. Under suitable conditions this also may become audible as a "summation tone" (p. 234).

It must be said at once that this rough-and-ready explanation of difference and summation tones is incomplete, and in a sense inaccurate. It is only inserted here to give a preliminary glimpse of an important problem which will be discussed more fully below (p. 231).

Forced Vibrations

We pass on now to an experiment of a different kind. We take a tuning-fork making (say) 261 vibrations a second, and fix securely between its two prongs a magnet like that already used in our telephone receiver (fig. 2), this being wound round with a wire through which an alternating current can be passed. Each change of the current in the wire changes the magnetisation of the soft iron, and this now changes the pull on the prongs of the tuning-fork.

Thus the current pulls the prongs of the fork about to follow its own changes, just as it pulled the diaphragm about on p. 9.

If the fork is stroked with a violin-bow, we shall hear the note c' produced by its 261 vibrations per second. While the vibration is still in progress, let us pass through the wire an alternating current of some other frequency, say 293 cycles a second. At first we hear a confused discord of sounds, but very shortly the fork will again be sounding a pure note. This will no longer be its own note c', but d', of which the frequency is 293.

It is easy to see what has happened. The current, pulling the fork about to follow its own changes, compels it to vibrate 293 times a second. At first this vibration is superposed on to the natural vibrations of the fork, of frequency 261, and produces 32 beats a second—an unpleasant number. As there is nothing at work to maintain the original vibrations of the fork, these soon die down; on the other hand, the current and magnet continually supply energy to the vibrations of frequency 293, so that these are maintained at full strength. After a time they are left alone in the field, and the fork, although tuned to emit the note c', is heard emitting the note d'. A vibration of this kind is known as a "forced" vibration, while

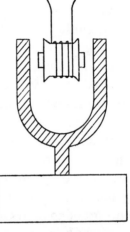

Fig. 21. Electrically driven tuning-fork. Instead of the fork being excited by striking or bowing, its prongs are kept in continuous vibration by a magnet and coil placed between them.

the vibration which a vibrator executes when it is set into motion and left to itself is called a "free" vibration.

Our experiment has illustrated a general principle of physics, which is as follows. When any vibrator is acted on by a regular periodic force of definite frequency, it may at first emit the note corresponding to the frequency of its free vibrations, but will soon settle down and emit the note whose frequency is that of the forcing agency. The free vibrations are transitory, but the forced vibrations are permanent.

It is easy to find applications of this principle. Our wireless sets have any number of free vibrations—we can set them all going by hitting the set in various places with a hammer—yet when we pass the ordinary alternating current of 50 cycles a second through the set, we only hear a steady hum, whose pitch corresponds to a frequency of 50. In the same way the telephone diaphragm which we employed in our first experiment (p. 9) has its own free vibrations, yet it only emits the musical note or sound which is being conveyed by the current in the wire.

To extend our experiment, let us suppose that the frequency of our alternating current can be altered at will. It has so far stood at 293; let us gradually lower it. As it drops, the pitch of the note we hear drops with it, passing from d' through c♯', c', B, B♭, and so on. This is not remarkable, but another feature is. For as the pitch falls, we hear the note getting steadily louder, until by the time it has reached c', the pitch of the free vibrations of the fork, it has become intensely loud. After it has passed beyond c', it becomes continually softer until finally it is inaudible.

Resonance

This last experiment provides a simple instance of a very general physical principle, known as the "Principle of Resonance". This may be stated as follows: the amplitude of a forced vibration increases as the period of the vibration approaches that of a free vibration of the vibrating system, and becomes very large when the two periods exactly coincide.

We shall best understand the meaning of the principle by studying a simple application of it—the ringing of a heavy church-bell.

A big bell is usually so heavy that the bell-ringer cannot hope to set it ringing by a single pull on his rope. He accordingly pulls the bell-rope with the strongest pull he can comfortably manage, and then lets go. The bell then begins to swing, somewhat like a pendulum in simple harmonic motion and, as it swings, the bell-rope moves up and down. By the time the bell has performed a complete swing, the rope has returned to its original position, and is moving downwards. The bell-ringer now gives it another pull, thereby superposing a new simple harmonic motion in the same phase as the original motion, and so causing the bell to swing more violently. The ringer repeats the process time after time. In so doing, he is in effect applying to the bell a periodic force, of period exactly equal to that of the free oscillations of the bell. This superposes new oscillations that are all in the same phase as the original oscillation, and so repeatedly increase the amplitude of this oscillation, until the bell is ringing with as much vigour as the bell-ringer desires. The effect is much the same if the ringer pulls on the rope at any instant whatever of its downward motion; on the other hand, if he were to pull on the rope during its upward motion, he would be adding a new oscillation in such a phase as to diminish the amplitude of the original oscillation, and by a repetition of this process he could bring the bell to rest.

Another instance of resonance is provided by the rolling of a ship in a cross sea. Each wave that strikes against the ship's side sets up a roll, and if the waves come at all regularly they may "force" a roll of considerable amplitude. The roll will have the same period as the waves, and if this period chances to coincide with the free period of the ship for rolling, the situation may become one of great danger.

We find another example of the principle of resonance in the suspension bridge. A suspension bridge is capable of swinging to and fro like a pendulum, so that it forms a vibrating structure, and may happen to have a free period near to that of a man's step. If so, a man walking across the bridge with a steady step may set up forced oscillations of quite large amplitude, while a body of men marching in step may set up oscillations violent enough to endanger the safety of the bridge. As there are cases on record of bridges being destroyed in this way, troops are ordered to break step when they come to a suspension bridge.

Again, a tumbler or wine-glass has its own very definite periods of free vibration, for on drawing a wet finger round the rim we hear a clear musical note, the pitch of which can be varied by putting more or less water into the glass. A singer can set the glass into vibration by singing this particular note near to it, and the more perfectly his voice is in tune with the glass, the more violent its vibrations will be. If he can sing loudly and truly enough, he may be able to shatter the glass into fragments.

The same effect may be noticed in a less agreeable form when a musical instrument sets furniture, ornaments and windows in a room into vibration. A particular note of a piano may often cause a disagreeable jangle of sound in one particular part of the room; we can usually trace this to some object which has a free vibration of the same frequency as the note in question.

To take yet another instance, the diaphragm of the loudspeaker of a radio set has its own frequencies of free vibration, and care must be taken that these shall not coincide with the frequencies of any of the sounds to be transmitted—otherwise these sounds would be heard in intolerable strength, as of course they sometimes are. Similar remarks apply to the telephone receiver. In both these cases very drastic measures are taken to ensure that no particular tones shall be unduly reinforced by resonance (see p. 240, below).

These examples may have seemed to suggest that the principle of resonance is a great nuisance, as indeed it often is. At the same time, it can be of very great assistance in solving the practical problems of science. If we blow across the open end of a key or a metal tube, we hear a distinct musical note, the pitch of which indicates the frequency of the free vibrations of the air inside the tube. Any musical note of this same frequency will accordingly set the air inside the tube into vibration by resonance. Helmholtz used to employ a number of glass vessels of such sizes and shapes that their free vibrations had the same frequencies as the notes of the musical scale. Each vessel had a small tube or spout protruding, and by placing this into or close to the ear, it was possible to tell whether the air inside was in vibration or not, so that the pitch of a musical note could be determined by observing which vibrator was set into resonance by it. Such vessels are commonly known as "Helmholtz resonators".

The wires of a piano can be used for the same purpose. If we raise the dampers by depressing the pedal, each string becomes a resonator, and will be set into vibration whenever a note is sounded whose pitch coincides with, or is near to, its own. The more closely the two pitches coincide, the more violent the vibrations will be. The vibrations can be made visible by putting tiny shavings of wood bent to a Λ-shape astride across the wires. When any wire is set into vibration, the chips lying on it begin to dance about, and quite moderate vibrations will unseat them altogether. Or, better still, we may lay fine unbent chips across the wires in directions at right angles to their lengths. When the wires of any note are set into even slight vibration, the chips lying across these particular wires begin to turn round and finally drop between the wires. The constituents of a chord, or other composite sound, can be discovered by placing fine chips over all the wires and noticing which of them fall through.

Sound Analysis

Instruments have recently been designed which analyse complex sounds with far more sensitiveness and accuracy than is possible with the simple methods just described.

The action of such instruments is best explained through their analogy with an ordinary radio receiving set. If the set is switched on, and the tuning-knob turned through the whole range of wave-lengths, we hear alternations of sound and silence. We may, for instance, hear loud sounds when the tuning-dial indicates wave-lengths of 1500 metres (200 kilocycles) or of 342·1 metres (879 kilocycles), fainter sounds at various other wave-lengths, and so on.

The reason for this is as follows. The set contains an electric oscillator, and the frequency of the free vibrations of this oscillator is not fixed, but changes when the tuning-knob is turned. When the tuning-dial shews 1500 metres or 200 kilocycles, the oscillator has a frequency of 200,000 cycles a second, and so is in perfect resonance with any electromagnetic waves of frequency 200,000 cycles a second which may be falling on the aerial. Because of this resonance, the sounds carried by waves of this particular frequency are heard loud and clear. Those carried by other waves are not heard because these waves are not in resonance with the oscillator, and so do not set up appreciable oscillations in it; their turn comes when the tuning-dial points elsewhere. Used in this way the set can be made to analyse the electromagnetic waves which fall on its aerial, sorting them out according to their frequencies and informing us as to their strength.

In the same way the modern sound analyser contains a sound resonator, the frequency of which can be varied continuously. If it is run through its range of frequencies in the presence of a composite sound, the air inside it will be set in agitation at some frequencies and not at others. The former frequencies must then be present in the sound, the latter not. The strengths of the various components can be recorded by electrical means.

Finally, the principle of resonance provides us with the means of sustaining a pure musical tone for as long as we wish. The apparatus is known as an "electrically maintained" tuning-fork, and is simply that which has been already described on p. 53, a fork in which vibrations can be "forced" by an intermittent or alternating current. On making the period of the current coincide with that of the fork, the latter gives out a loud pure note which can be prolonged indefinitely.

CHAPTER III

THE VIBRATIONS OF STRINGS AND HARMONICS

We began our study of sound-curves by examining the curve produced by a tuning-fork. A tuning-fork was chosen, because it emits a perfectly pure tone. But, as every musician knows, its sound is not only perfectly pure, but is also perfectly uninteresting to a musical ear—just because it is so pure.

The artistic eye does not find pleasure in the simple figures of the geometer—the straight line, the triangle or the circle—but rather in a subtle blend of these in which the separate ingredients can hardly be distinguished. In the same way, the painter finds but little interest in the pure colours of his paint-box; his real interest lies in creating subtle, rich or delicate blends of these. It is the same in music; our ears do not find pleasure in the simple tones we have so far been studying but in intricate blends of these. The various musical instruments provide us with ready-made blends, which we can combine still further at our discretion.

In the present chapter we shall consider the sounds which are emitted by stretched strings—such as, for instance, are employed in the piano, violin, harp, zither and guitar—and we shall find how to interpret these as blends of the pure tones we have already had under consideration.

Experiments with the Monochord

Our source of sound will no longer be a tuning-fork but an instrument which was known to the ancient Greek mathematicians, Pythagoras in particular, and is still to be found in every acoustical laboratory—the monochord.

Its essentials are shewn in fig. 22. A wire, with one end A fastened rigidly to a solid framework of wood, passes over a fixed bridge B and a movable bridge C, after which it

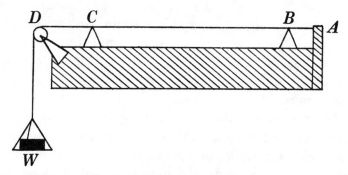

Fig. 22. The monochord. The string is kept in a state of tension by the suspended weight W, while "bridges" like those of a violin limit the vibration to a range BC. The instrument is arranged so that both range and tension are under control.

passes over a freely turning wheel D, its other end supporting a weight W. This weight of course keeps the wire in a state of tension, and we can make the tension as large or small as we please by altering the weight. Only the piece BC of the string is set into vibration, and as the bridge C can be moved backwards and forwards, this can be made of any length we please. It can be set in vibration in a variety of ways—by striking it, as in the piano; by stroking it with a bow, as in the violin; by plucking it, as in the harp; possibly even by blowing over it as in the Aeolian harp, or as the wind makes the telegraph wires whistle on a cold windy day.

On setting the string vibrating in any of these ways, we hear a musical note of definite pitch. While this is still sounding, let us press with our hand on the weight W. We shall find that the note rises in pitch, and the harder we press on the weight, the greater the rise will be. The pressure of our hand has of course increased the tension in the string, so that we learn that increasing the tension of a string raises the pitch of the note it emits. This is the way in which the violinist and piano-tuner tune their strings and wires; when one of these is too low in pitch, they screw up the tuning-key.

A series of experiments will disclose the exact relation between the pitch and the tension of a string. Suppose that the string originally sounds c′(middle C), the tension being 10 lb. To raise the note an octave, to c″, we shall find we must increase the tension to 40 lb.; to raise it yet another octave to c‴, we need a total tension of 160 lb., and so on. In each case a fourfold increase in the tension is needed to double the frequency of the note sounded, and we shall find that this is always the case. It is a general law that the frequency is proportional to the square root of the tension.

We can also experiment on the effect of changing the length of our string, repeating experiments such as were performed by Pythagoras some 2500 years ago. Sliding the bridge C in fig. 22 to the right shortens the effective length BC of the string, but leaves the tension the same—that necessary to support the weight W. When we shorten the string, we find that the pitch of the sound rises. If we halve its length, the pitch rises exactly an octave, shewing that the period of vibration has also been halved. By experimenting with the bridge C in all sorts of positions, we discover the general law that the period is exactly proportional to the length of the string, so that the frequency of vibration varies inversely as the length of the string. This law is exemplified in all stringed instruments. In the violin, the same string is made to give out different notes by altering its effective length by touching it with the finger. In the pianoforte different notes are obtained from wires of different lengths.

We may experiment in the same way on the effect of changing the thickness or the material of our wire.

Mersenne's Laws

The knowledge gained from all these experiments can be summed up in the following laws, which were first formulated by the French mathematician Mersenne (*Harmonie Universelle*, 1636):

I. When a string and its tension remain unaltered, but the length is varied, the period of vibration is proportional to the length. (The law of Pythagoras.)

II. When a string and its length remain unaltered, but the tension is varied, the frequency of vibration is proportional to the square root of the tension.

III. For different strings of the same length and tension, the period of vibration is proportional to the square root of the weight of the string.

The operation of all these laws is illustrated in the ordinary pianoforte. The piano-maker could obtain any range of frequencies he wanted by using strings of different lengths but similar structure, the material and tension being the same in all. But the $7\frac{1}{4}$ octaves range of the modern pianoforte contains notes whose frequencies range from 27 to 4096. If the piano-maker relied on the law of Pythagoras alone, his longest string would have to be more than 150 times the length of his shortest, so that either the former would be inconveniently long, or the latter inconveniently short. He accordingly avails himself of the two other laws of Mersenne. He avoids undue length of his bass strings by increasing their weight—usually by twisting thinner copper wire spirally round them. He avoids inconvenient shortness of his treble strings by increasing their tension. This had to be done with caution in the old wooden-frame piano, since the combined tension of more than 200 stretched strings imposed a great strain on a wooden structure. The modern steel frame can, however, support a total tension of about 30 tons with safety, so that piano-wires can now be screwed up to tensions which were formerly quite impracticable.

The Free Vibrations of a String

Having found or verified these laws, we may pass to a slightly more complicated experiment, known as Melde's experiment. We remove the bridge B and the fastening at A altogether, and replace our wire by a fine string or thread of silk. One end of this still passes over the wheel D and supports the weight W, but its other end is fastened to one prong of a tuning-fork, so that the apparatus now looks like that sketched in fig. 23.

We can set the length AC of the string into vibration by bowing it with a violin-bow, in which case it will of course execute its own free vibrations, and we shall hear a faint musical note—the note of the string. But we can also set the string into vibration by bowing the tuning-fork. This now performs its own free vibrations, and transmits its motion to the string, so that this will execute "forced" (p. 52) vibrations of the same frequency as those of the fork. We now hear the note of the fork, and this will usually be different from that of the string previously heard. If, however, the two happen to be of the same pitch, there will

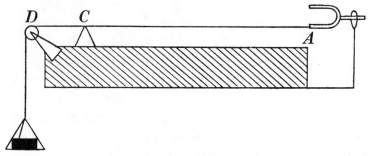

Fig. 23. The monochord arranged for Melde's experiment.

be resonance (p. 55) between the fork and the string. The vibrations of the latter will now be exceptionally violent, and visible to the eye.

If we experiment with a whole series of forks, or with an electrically driven fork of which the pitch can be varied, we shall find that there is not only one pitch of fork, but a whole series of pitches, for which the vibrations of the string become so violent as to be visible. At each of these pitches, there must be resonance between the fork and string, so that the string must have a free vibration corresponding to each; the experiment discloses the frequencies of the free vibrations of the string.

We find that the frequencies of these free vibrations are all multiples of the same number. For instance, if the deepest pitched fork which produces resonance has a frequency of 256, the frequencies of the other forks will be 512, 768, 1024, etc., these being respectively twice, three times, four times, etc. the original number 256.

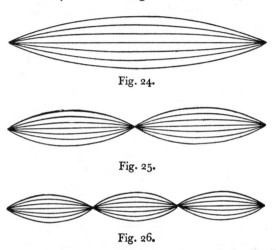

Fig. 24.

Fig. 25.

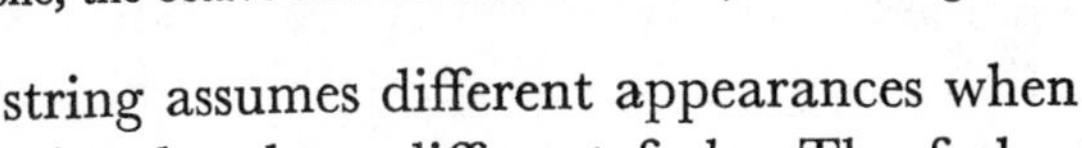

Fig. 26.

Figs. 24, 25 and 26. Characteristic vibrations of a stretched string. The string vibrates in one, two and three equal parts respectively, and emits its fundamental tone, the octave and the twelfth of this, in so doing.

The string assumes different appearances when it is set in vibration by these different forks. The fork of lowest pitch, of frequency 256, causes it to look as in fig. 24; the next, of frequency 512, gives it the appearance shewn in fig. 25; that of frequency 768 the appearance of fig. 26, and so on. The fork whose frequency is thrice that of the fork of lowest pitch makes the string vibrate in three separate equal parts, and so on throughout. Thus the periods of the various free vibrations are in every case proportional to the lengths of the string which are in separate vibration, in accordance with the law of Pythagoras.

However often we perform our experiments, we shall find no free vibrations other than these, and it is easy to see that there can be no others.

For suppose that the string could vibrate separately in two parts which formed two-fifths and three-fifths respectively of the whole, as in fig. 27. Then the periods of these two vibrations would be proportional to the lengths of the

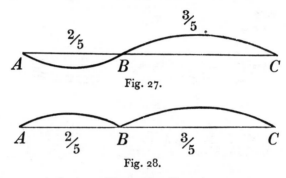

Fig. 27.

Fig. 28.

Figs. 27 and 28. A possible position (fig. 27) and an impossible position (fig. 28) for a vibrating string.

separate parts, by the law of Pythagoras, and so would be in the ratio of 2 : 3. After the part BC of the string had performed a complete vibration, it would be back in the position shewn in fig. 27, but at this instant the part AB would have performed $1\frac{1}{2}$ vibrations, and so would be in the position shewn in fig. 28. We see at once that this is an impossible position for a string which is fixed only at A and C. It would be possible if the string were fixed also at B, but then there would be a pull on the fastening at B. Without such a pull, the point B of the string would have been drawn upwards before the string had reached the configuration shewn in fig. 28, so that the string cannot possibly be in such a configuration.

This only proves the result for one particular case, but it is easy to see that the principle is universally true. To avoid an impossible kink like that shewn in fig. 28, the vibrations of the different segments of the string must keep exactly in step, and so must have precisely the same period. This means that the separate vibrating segments must all be of the same length, which limits us to the vibrations we have already discovered.

Waves travelling along a String

Let us now repeat the experiment in a somewhat different form, replacing the thread by a more massive string or rope, and the tuning-fork by our hand. We hold the free end of

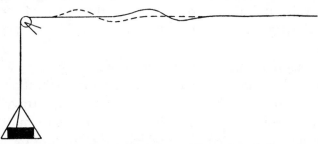

Fig. 29. The passage of a single wave along a stretched string.

the string in our hand, and very slowly raise it. The whole string of course follows and assumes a slanting position; we slowly lower our hand again, and the string resumes its horizontal position. Suppose, however, that we now speed up our tempo, and raise and lower our hand by rapid jerks. The inertia of the string now comes into play, with the result that the first motion sends a sort of wave of elevation along the string, while the second sends a wave of depression which follows the former at a short interval—at one instant the rope may stand as in the thick line in fig. 29; at a later instant it will lie as shewn by the dotted line. As each wave reaches the wheel, it is reflected back along the string. A wave travels along the string much as a ripple travels along the surface of a pond, and is reflected at the immovable end much as the ripple is reflected by a stone wall.

We are now able to see what happened in Melde's experiment, which we have just performed. In place of our hand we had the prong of a tuning-fork, and instead of moving up and down only once, this moved up and down repeatedly and regularly. It did not send one single wave travelling along the wire, but a whole succession at regular intervals, and these, together with the reflected waves, must have given the motion we observed in Melde's experiment.

This latter was, however, one or other of the free vibrations of the string, whence we see that a free vibration can be represented as the superposition of a series of travelling waves; these travel in both directions, because we must not overlook the reflected waves which travel towards the fork. Fig. 30 shews how travelling waves combine to give the free vibrations.

Thus we may regard the motion of a string in two alternative ways: either as being made up of a number of free vibrations, or as made up of a number of travelling waves; these are merely two ways of looking at the same thing.

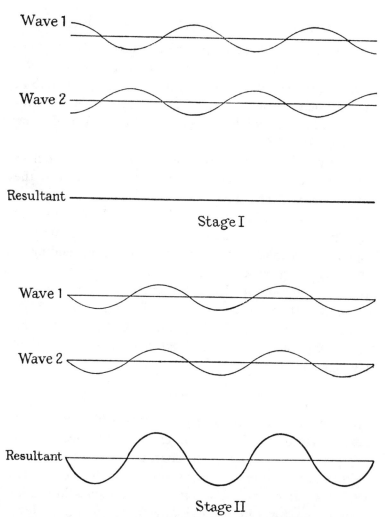

Fig. 30. A free vibration as the equivalent of two travelling waves. Wave 1 travels to the left, and wave 2 to the right. In stage I the waves are in exactly opposite phase and so neutralise one another, the resultant displacement being *nil* all along the string. After wave 1 has travelled a quarter wave-length to the left, and wave 2 a quarter wave-length to the right, stage II is reached. The waves are now exactly in the same phase, and so reinforce one another to the fullest possible extent, their resultant being as shewn. As the motion proceeds further, the phase difference increases, until it finally reaches 180 degrees, and the motion is back at stage I.

When we regard the motion in the second way, namely as travelling waves, the law of Pythagoras—that the period of a free vibration is proportional to the length of string which vibrates—admits of a very simple interpretation. The law merely asserts that all waves travel along the string at precisely the same speed.

The two other laws of Mersenne (on p. 64) now tell us that the speed at which the waves travel varies as the square root of the tension, and inversely as the square root of the weight, of the string. It is, indeed, obvious that something of this sort must be true. For if we increase the tension in the string, we increase the forces tending to move the string back to its original position, and so speed up the whole motion. If, on the other hand, we increase the weight of the string, we increase the mass to be moved by these forces, and so slow down the whole motion. Similar laws are true, *mutatis mutandis*, for every kind of wave-motion.

Harmonics

We have seen that, when any structure whatever is in a state of vibration, its motion can be regarded as the superposition of a number of separate free vibrations.

When any one of these vibrations is performed alone, every particle of the vibrating structure moves in simple harmonic motion—in the same way, that is, as a particle of our tuning-fork. It follows that if musical sounds are produced by the different vibrations, these must be all pure tones of the kind discussed on p. 17; in other words, each could be produced by the free vibrations of a tuning-fork.

In the special case we have just been discussing, in which the vibrating structure is a stretched string, the frequencies of the different vibrations stand in the simple ratio $1 : 2 : 3 : 4 : \ldots$, so that the whole sound produced by the vibrations of the string could be produced by the simultaneous vibrations of a whole battery of tuning-forks, whose frequencies were in the ratio $1 : 2 : 3 : 4 : \ldots$. We begin to see why the sound of a violin or a piano is richer, and so more interesting, than that of a tuning-fork.

Theoretical considerations have shewn us that tones of all these frequencies and no others will be sounded when our string is played, and by using a set of Helmholtz resonators or a sound analyser of the kind described on p. 59, we can verify that this is the case. For instance, if a violin is made to sound c′ (middle C) of frequency 256, resonators of frequencies 256, 512, 768, 1024, 1280, etc. will be set into vibration by resonance, the others not.

Or we may use the strings of a piano as resonators (p. 58). We raise the dampers by depressing the pedal and play on a violin the note c′, tuned to the c′ of the piano. We naturally find that the piano-wires of the note c′ are set into vibration by resonance, but we shall find that the wires of the following notes are also vibrating: c″, g″, c‴, e‴, g‴, bb‴, c^iv, etc. Referring to the table on p. 27, we find that these notes have the frequencies given in the second line of the table below, the frequencies of the vibrations of the violin-string being those given in the third line:

Piano note	c′	c″	g″	c‴	e‴	g‴	bb‴	c^iv
Frequency	256	512	767	1024	1290	1534	1825	2048
Frequency of string	256	512	768	1024	1280	1536	1792	2048
Number of harmonic	1	2	3	4	5	6	7	8

Of the various tones sounded by the vibrating string, that of lowest frequency—256 in the present case—is called the "foundation tone" or "fundamental tone" of the string, while the whole series of pure tones which are blended in the sound of the string are called "harmonics". The fundamental tone is described as the first harmonic, the tone of the frequency next above is the second harmonic, and so on.

The table on p. 73 shews that the second harmonic is the octave of the fundamental, while the fourth harmonic is the super-octave or fifteenth. The third harmonic does not coincide exactly with the twelfth g″—and herefrom arises a complication which will require much discussion in a subsequent chapter—but for our present purpose it is near enough, and we may say that the third harmonic is the twelfth of the fundamental note. To the same degree of approximation the fifth harmonic is the seventeenth, the sixth harmonic is the nineteenth, and so on.

Using the piano as a resonator, we can verify not only the frequencies of these various harmonics, but also the nature of the vibrations by which they are produced. We have already seen, for instance, that when we strike c′ on the piano, the strings of c″, g″, c‴, etc. are all set vibrating through resonance, shewing that the c′ strings have free vibrations with the frequencies of all these higher tones. Let us now reverse the experiment, and strike c″ on the piano. As the c′ strings have free vibrations of the same frequency as the note now sounding, they will be set in vibration, and if we have laid small chips of wood across them in the way already explained, these chips will fall off. But if one of these small chips happens to be just half-way along the string, this particular chip will not fall off. All the others fall off, but this does not, shewing that the string is vibrating everywhere except here. In fact the string has exactly the motion shewn in fig. 25 (p. 67); it is vibrating in two equal parts. In the same way we can verify that striking g″ causes the strings of c′ to vibrate in three equal parts in the way shewn in fig. 26, and so on indefinitely.

Nodes and Loops

If we could take an instantaneous photograph of a string while it was performing any one of its free vibrations alone, we should find that its shape at any instant is a "sine curve" or a "simple harmonic curve" of the kind described on p. 28. It can, indeed, be proved by rigorous mathematics that this must be the case, provided only that the string is perfectly flexible and is formed of uniform material throughout.

The distance by which any particle of the string is displaced from its normal position P, such as PQ in fig. 31,

Fig. 31. The curve formed by a string performing one vibration alone is always a simple harmonic curve.

is called the displacement at P. As we pass along the string, the amount of this displacement continually waxes and wanes. The points at which it is zero, i.e. where the string does not move at all, are known as nodes. Obviously the ends of the string, or any other points at which it is securely fastened, must be nodes (nodus = a knot), and the remaining nodes, as we have seen, lie evenly spaced between these two end nodes. The points at which the displacement is a maximum are known as antinodes or loops; these lie evenly spaced between the nodes. Touching the string at a node of its vibration produces no effect, because the string has no motion there; if, however, we touch it at any point except a node, we kill the vibration. If it is not merely performing a single vibration, but vibrating in any other way, a touch will kill all the vibrations except those for which the point of contact is a node. For instance, if we touch the c′ string of a piano at its middle point while it is vibrating, we silence the tones c′, g″, e‴, etc., but leave it sounding c″, c‴, g‴, etc., because the middle point of the string is a node for all even-numbered harmonics. Violinists are accustomed to produce harmonics in a similar way, by lightly touching the string at the appropriate nodes.

String Tone

We have now seen that when we play c′ on the violin or piano we are in effect playing the chord of pure tones which is shewn on the right of fig. 32; this will obviously

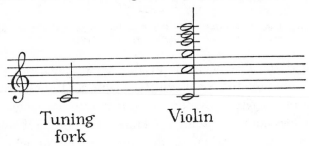

Fig. 32. The tuning-fork only sounds single pure tones. A single note of the violin, on the other hand, sounds a whole chord of pure tones—hence the richness and fullness of good violin tone.

sound fuller, brighter and richer than the pure note of the tuning-fork shewn on the left.

In this diagram, all the harmonics are shewn in equal strength, but in practice the fundamental tone and the lower harmonics are frequently much stronger than the others, these becoming progressively weaker as we ascend. Theoretically the chord ascends to infinity; often in practice harmonics beyond the sixth or seventh are too weak to affect the ear, so that the note is as shewn above. From the musical point of view, it is fortunate that this is the case, for many of the higher harmonics—including all odd-numbered harmonics above the fifth—form a dissonance with the fundamental note.

Fig. 32 refers only to the vibrations of one special structure—the stretched string. Other structures also have a great number of distinct free vibrations, but their frequencies are not connected by the simple numerical relations we have just found. This simplicity is peculiar to the stretched string, although the vibrations of a column of air, as in a flute or organ-pipe, may approach very near to it. The vibrations of stretched membranes, as in drums, and of solid structures such as bells, cymbals and triangles, usually shew no simple relations between their frequencies, so that the sounds they emit are generally discordant musically. An instance of this is provided by the clang tones of a tuning-fork (p. 33).

The discussion which now follows will make it clear why the vibrations of a stretched string are of such a specially simple and particularly musical nature.

Harmonic Analysis

Several times already we have superposed two simple harmonic curves, and studied the new curves resulting from the superposition. The essence of the process of superposition has already been illustrated in fig. 12 *a* on p. 33, and fig. 17 on p. 39. In each of these cases the number of superposed curves is only two; when a greater number of such curves is superposed, the resultant curve may be of a highly complicated form.

There is a branch of mathematics known as "harmonic analysis" which deals with the converse problem of sorting out the resultant curve into its constituents. Superposing a number of curves is as simple as mixing chemicals in a test-tube; anyone can do it. But to take the final mixture and discover what ingredients have gone into its composition may require great skill.

Fortunately the problem is easier for the mathematician than for the analytical chemist. There is a very simple technique for analysing any curve, no matter how complicated it may be, into its constituent simple harmonic curves. It is based on a mathematical theorem known as Fourier's theorem, after its discoverer, the famous French mathematician J. B. J. Fourier (1768–1830),

The theorem tells us that every curve, no matter what its nature may be, or in what way it was originally obtained, can be exactly reproduced by superposing a sufficient number of simple harmonic curves—in brief, every curve can be built up by piling up waves.

The theorem further tells us that we need only use waves of certain specified lengths. If, for instance, the original curve repeats itself regularly at intervals of one foot, we need only employ curves which repeat themselves regularly 1, 2, 3, 4, etc. times every foot—i.e. waves of lengths 12, 6, 4, 3, etc. inches. This is almost obvious, for waves of other lengths, such as 18 or 5 inches, would prevent the composite curve repeating regularly every foot. If the original curve does not repeat regularly, we treat its whole length as the first half-period* of a curve which does repeat, and obtain the theorem in its more usual form. It tells us that the original curve can be built up out of simple harmonic constituents such that the first has one complete half-wave within the range of the original curve, the second has two complete half-waves, the third has three, and so on; constituents which contain fractional parts of half-waves need not be employed at all. There is a fairly simple rule for calculating the amplitudes of the various constituents, but this lies beyond the scope of the present book.

We obtain a first glimpse into the way of using this theorem if we suppose our original curve to be the curve assumed by a stretched string at any instant of its vibration. Figs. 24, 25 and 26 on p. 67 shew groups of simple harmonic curves which contain one, two and three complete half-waves respectively within the range of the string. Let us imagine this series of diagrams extended indefinitely so as to exhibit further simple harmonic curves containing 4, 5, 6, 7 and all other numbers of complete half-waves. Then the series of curves obtained in this way is precisely the

---

* It might seem simpler to treat the original curve as a whole period of a repeating curve, but there are mathematical reasons against this.

series of constituent curves required by the theorem. We take one curve out of each diagram, and superpose them all; the theorem tells us that by a suitable choice of these curves, the final resultant curve can be made to agree with any curve we happen to have before us. Or, to state it the other way round, any curve we please can be analysed into constituent curves, one of which will be taken from fig. 24, one from fig. 25, one from fig. 26, and so on.

This is not, of course, the only way in which a curve can be decomposed into a number of other curves. Indeed, the number of ways is infinite, just as there is an infinite number of ways in which a piece of paper can be torn into smaller pieces. But the way just mentioned is unique in one respect, and this makes it of the utmost importance in the theory of music. For when we decompose the curve of a vibrating string into simple harmonic curves in this particular way, we are in effect decomposing the motion of the string into its separate free vibrations, and these represent the constituent tones in the note sounded by the vibration. As the vibratory motion proceeds, each of these free vibrations persists without any change of strength, apart from the gradual dying away already explained. If, on the other hand, we had decomposed the vibration in any other way, the strength of the constituent vibrations would be continually changing —probably hundreds of times a second—and so would have no reference to the musical quality of the sound produced by the main vibration.

So general a theory as this may well seem confused and highly complicated, but a single detailed illustration will bring it into sharp focus and shew its importance.

String Plucked at its Middle Point

Let us displace the middle point of a stretched string AB to C, so that the string forms a flat triangle ACB as in fig. 33.

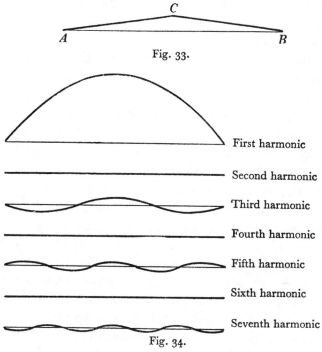

Fig. 33.

First harmonic

Second harmonic

Third harmonic

Fourth harmonic

Fifth harmonic

Sixth harmonic

Seventh harmonic

Fig. 34.

Figs. 33 and 34. The string is displaced to form the triangle ACB. This "curve" can be analysed into the simple harmonic curves shewn in fig. 34. On superposing these we restore the "curve" ACB of fig. 33. (The vertical scales in fig. 34 are all magnified ten-fold.)

The shape of the string ACB may still be regarded as a curve, although a somewhat unusual one, and our theorem tells us that this "curve" can be obtained from the superposition of a number of simple harmonic curves. In actual fact, fig. 34 shews how the curve *ACB* can be resolved into its constituent curves; if we superpose all the curves shewn in this latter figure, we shall find we have restored our original broken line *ACB*, except for a difference in scale; the vertical scale in fig. 34 has been made ten times the horizontal in order that the fluctuations of the higher harmonics may be the more clearly seen.

Suppose we now let go of the point *C*, and allow the natural motion of the string to proceed. We may imagine each of the curves shewn in fig. 34 to decrease and increase rhythmically in its own proper period in the way described on p. 71, and the superposition of the curves at any instant will give us the shape of the string at that instant. These curves correspond to the various harmonics that are sounded on plucking a string at its middle point.

We notice that the second, fourth and sixth harmonics are absent. This is not a general property of harmonics, but is peculiar to the special case we have chosen. We have plucked the string in such a way that its two halves are bound to move in similar fashion, and as a consequence the second, fourth and sixth harmonics, which necessarily imply dissimilarity in the two halves, cannot possibly appear. If we had plucked it anywhere else than at its middle point, some at least of these harmonics would have been present.

Analysis of a Sound-Curve

Let us next apply Fourier's theorem to a piece of a sound-curve. The theorem tells us that any sound-curve whatever can be reproduced by the superposition of suitably chosen simple harmonic waves. Consequently any sound, no matter how complex—whether the voice of a singer or a motor-bus changing gear—can be analysed into pure tones and reproduced exactly by a battery of tuning-forks, or other sources of pure tone. Professor Dayton Miller has built up groups of organ-pipes, which produce the various vowels when sounded in unison; other groups say *papa* and *mama*.

The sound-curve of a musical sound is periodic; it recurs at perfectly regular intervals. Indeed, we have seen that this is the quality which distinguishes music from noise. Fourier's theorem tells us that such a sound-curve can be made up by the superposition of simple harmonic curves such that 1, 2, 3, or some other integral number of complete waves occur within each period of the original curve. If, for instance, the sound-curve has a frequency of 100, it can be reproduced by the superposition of simple harmonic curves of frequencies 100, 200, 300, etc.

Each of these curves represents a pure tone, whence we see that any musical sound of frequency 100 is made up of pure tones having respectively 1, 2, 3, etc. times the frequency of the original sound. These tones are called the "natural harmonics" of the note in question.

Natural Harmonics and Resonance

Vibrations are often set up in a vibrating structure by a force or disturbance which continually varies in strength; such a force may be periodic in the sense that the variations repeat themselves at regular intervals. Fourier's theorem now tells us that a variable force of this kind can be resolved into a number of constituent forces each of which varies in a simple harmonic manner, and that the frequencies of these forces will be 1, 2, 3 ... times that of the total force. For instance, if the force repeats itself 100 times a second, the simple harmonic constituents of the force will repeat themselves 100, 200, 300, etc., times a second.

If the structure has free vibrations of frequencies 100, 200, 300, etc., these will be set vibrating strongly by resonance, while any vibrations of other frequencies that the structure may possess will not be set going in any appreciable strength. In other words, a disturbing force only excites by resonance the "natural harmonics" of a tone of the same period as itself.

This, as we shall see later, is a result of great importance to music in general. Amongst other things, it explains why the stretched string has such outstanding musical qualities; the reason is simply that its free vibrations coincide exactly in frequency with the natural harmonics of its fundamental tone, so that when the fundamental tone is set going, the harmonics are set going as well.

Timbre and the Harmonic Analysis of Sound

By *timbre* is meant the distinguishing or characteristic quality of a sound; it is by their timbre that we recognise an instrument, a voice or the quality of an organ-stop, regardless of the pitch or intensity of the note it is sounding.

The investigations of Helmholtz proved that the timbre of a sound is determined by the proportions in which the various natural harmonics are heard in it. It is obvious that something of this kind must be true. We know, for instance, that the more we hear of the higher harmonics in any sound, the farther we get away from the dull quality of the tuning-fork, which is characterised by a complete absence of upper harmonics. Thus we may say that the upper harmonics add life, richness and interest to the foundation tone. And as they are all at least an octave higher in pitch, they will obviously add brilliance, and possibly shrillness also.

The detailed effects of the various harmonics are a matter for careful study. There are several devices which enable us to blend harmonics as we please, and study the result. On large organs, the choir manual frequently contains stops which sound the first eight (or even more) harmonics separately, and by combining these in various ways, sounds of different timbre can be produced, and their harmonic composition noted. The great manuals of old organs often contained similar selections of stops. There are also various electrical instruments which permit of the harmonics being blended in any relative strengths we desire.

In the well-known Hammond electric organ, pure tones are produced by alternating currents of the appropriate frequencies. Nine such tones can be produced on each manual key, these representing a selection of the harmonics of the normal note of the key or of its sub-octaves.* The instrument contains no stops in the ordinary sense, but the presence or absence of the various harmonics, as well as the proportions in which they enter, are controlled by drawbars, which can be drawn to any extent desired. Different blends of harmonics yield different qualities of tone, and

* Actually they are the fundamental tone, the second, third, fourth, fifth, sixth and eighth harmonics, together with the suboctave and the quint, the last being a harmonic of the 16-foot tone, but not, of course, of the 8-foot tone.

ready-made "prescriptions" for compounding various qualities of tone are supplied with the instrument. For instance, we are told to take

		For string tone	For open diapason tone	For clarinet tone	For reed tone
Sub-octave	to strength	0	0	0	5
Quint	„ „	0	0	0	5
Fundamental	„ „	1	7	6	5
Octave	„ „	4	7	4	5
Twelfth	„ „	5	5	7	5
Fifteenth	„ „	5	5	0	5
Seventeenth	„ „	5	2	5	5
Nineteenth	„ „	5	2	2	5
Twenty-second	„ „	5	0	0	5

In whatever way we experiment, we obtain results which are somewhat as follows:

The timbre depends only on the relative energies of the various harmonics and not on their phase-differences. Differences of phase produce no effect on the ear. This is known as Ohm's law, having been discovered by G. S. Ohm (1787–1854), the discoverer of the still better known electrical law.

The second harmonic adds clearness and brilliance but nothing else, it being a general principle that the addition of the octave can introduce no difference of timbre or characteristic musical quality. When the second harmonic is of equal strength with the first, it produces much the same effect as adding the octave-coupler on an organ or harmonium or playing in octaves, instead of single notes, on the piano.

The third harmonic again adds a certain amount of brilliance because of its high pitch, but it also introduces a difference of timbre, thickening the tone, and adding to it a certain hollow, throaty or nasal quality, which we may recognise as one of the main ingredients of clarinet tone (see opposite, and p. 150, below).

The fourth harmonic, being two octaves above the fundamental, adds yet more brilliance, and perhaps even shrillness, but nothing more, for the reason already explained. The fifth harmonic, apart from adding yet more brilliance, adds a rich, somewhat horn-like quality to the tone, while the sixth adds a delicate shrillness of nasal quality.

As the table on p. 73 shews, all these six harmonics form parts of the common chord of the fundamental note, and so are concordant with this note and with one another. The seventh harmonic, however, introduces an element of discord; if the fundamental note is c′, its pitch is approximately b♭‴, which forms a dissonance with c. The same is true of the ninth, eleventh, thirteenth, and all higher odd-numbered harmonics; these add dissonance as well as shrillness to the fundamental tone, and so introduce a roughness or harshness into the composite sound. The resultant quality of tone is often described as "metallic", since a piece of metal, when struck, emits a sound which is rich in discordant high tones.

Harmonic Synthesis

As the richness and quality of a musical note depend only on the proportions in which the different harmonics enter, it will be clear that the blending of harmonics to make a good musical tone is as important as the newspaper advertisements tell us that the blending of teas and tobaccos is. Let us then try to discover the art of making our musical instruments blend their own harmonics in such a way as to give us the particular quality of tone that our musical taste demands. For the moment we are concerned only with string tone, and the problem before us, stated in the simplest possible language, is to find out how to set a string vibrating in such a way as to encourage the harmonics we want, and to suppress the rest.

A result which has already been obtained provides a clue as to the line of attack on this problem.

We found that when a stretched string is plucked at its middle point, the second and fourth harmonics are absent from the sound produced, whereas if it is plucked at some other point, these harmonics are present.

The second and fourth are, however, the harmonics which above all others impart clearness and brilliance to the tone, so that the note given out by the plucked string will be deficient in these qualities. It will have a rather hollow, nasal quality, reminiscent perhaps of the tone of a clarinet or a stopped organ-pipe, since the tones of both of these consist mainly of odd-numbered harmonics.

This seems to suggest that the quality of tone emitted by a string depends on the point at which we pluck or strike the string, and harmonic analysis (p. 78) proves that this is so. It shews that the strength in which any particular harmonic occurs depends on the product of two distinct factors, which may be described as the position factor and the ordinal factor. We shall discuss these in turn.

The position factor depends solely on the position of the point at which we pluck or strike the string. Its value is very easily stated. Let us draw a diagram of the string performing the vibration corresponding to the harmonic in question, as in fig. 31 on p. 75. Then when the string is plucked or struck at any point P, the factor in question is simply PQ^2, the square of the displacement PQ. As we pass from a node to a loop, this factor increases steadily from zero to its maximum value. Thus we can obtain a harmonic in its fullest possible strength by exciting the string at a loop; we can eliminate a harmonic altogether by exciting the string at a node.

This is a most important result, of which we have already had an instance (on p. 82). The middle point of a string is a node for all even-numbered harmonics and a loop for all odd-numbered harmonics, so that if we excite a string at its middle point, all the even-numbered harmonics, including the octave, super-octave and all higher octaves, will be missing from the sound produced (the result already obtained), while all the odd-numbered harmonics will be present in their maximum strength. In the same way we see that if we excite the string at a point a third way along its length, the third harmonic will be missing, but the second (octave) and fourth (super-octave) will be fairly strong, giving a clear brilliant tone. If we excite the string a quarter way along, the second harmonic will be heard in full strength but the fourth will be entirely missing, while the third and fifth will appear, but weakly.

The value of the second or ordinal factor depends on which particular harmonic we have under consideration, and also on whether the string is plucked or struck.

Plucked String

If the string is plucked, as in a harp, guitar or harpsi-chord, the ordinal factor is very simple; it is proportional to the inverse square of the ordinal number of the harmonic in question. That is to say, if we assign the value 1 to the factor for the first harmonic, it will be $\frac{1}{4}$ for the second harmonic, $\frac{1}{9}$ for the third harmonic, $\frac{1}{16}$ for the fourth and so on. The factor for the seventh harmonic is $\frac{1}{49}$, which is but little more than 2 per cent., so that harmonics above the sixth contribute very little to the tone. We have already noticed that the first six harmonics—e.g. c′, c″, g″, c‴, e‴, g‴—are notes of the common chord, so that the sound of a plucked string will be fairly free from dissonant harmonics, and therefore almost entirely musical.

Struck String

If the string is struck with a hard sharp hammer, the ordinal factor is even simpler, for it has the same value for each harmonic. As there are an infinite number of harmonics, it might be thought that the total energy of the sound produced would have to be divided equally among this infinite number of harmonics, with the result that each separate harmonic would get no energy at all. Our mathematical theory has, however, been based upon the supposition that the string is perfectly flexible. No string is so in practice, and this want of perfect flexibility exerts a drag on the highest harmonics of all, and prevents them sounding in their full strength, or even in any appreciable strength at all. Even so, the number of harmonics which remain to share the energy in approximately equal shares is still very great, so that the share for each is very small. Thus the fundamental note and the few lower harmonics get but little of the total energy, the main bulk of this going into the many higher harmonics. As most of these are discordant both with the fundamental note and with one another, the result is a sharp shrill sound of metallic quality (p. 87)—it is, in fact, the noise we hear if we accidentally drop a key or a coin on to a piano-wire.

The following table exhibits the results just mentioned, and shews the contrast between the two types of sound we have so far considered:

Distribution of Energy between the various Harmonics of a String

Ordinal number of harmonic	1	2	3	4	5	6	7
Note	c′	c″	g″	c‴	e‴	g‴	b♭‴
Energy when plucked	1	$\frac{1}{4}$	$\frac{1}{9}$	$\frac{1}{16}$	$\frac{1}{25}$	$\frac{1}{36}$	$\frac{1}{49}$
Energy when struck	1	1	1	1	1	1	1

Ordinal number of harmonic	8	9	10	11	12	13	14
Note	c^{iv}	d^{iv}	e^{iv}	—	g^{iv}	—	b♭^{iv}
Energy when plucked	$\frac{1}{64}$	$\frac{1}{81}$	$\frac{1}{100}$	$\frac{1}{121}$	$\frac{1}{144}$	$\frac{1}{169}$	$\frac{1}{196}$
Energy when struck	1	1	1	1	1	1	1

Piano Tone

In the piano the wire is struck with a hammer covered with soft felt. The felt prolongs the duration of the impact, so that, by the time that the hammer finally breaks its contact with the string, a substantial length of the string has already been set in motion. This reduces the energy which goes into the higher harmonics, and so avoids the harsh jangle of sound represented in the bottom column of the above table. As we have seen that discord begins with the seventh harmonic, the hammer should be sufficiently felted to reduce the seventh and higher discordant harmonics (ninth, eleventh, etc.) to small proportions.

Even if the hammer were perfectly hard, the seventh harmonic could be eliminated entirely by allowing the hammer to strike the string at a point a seventh of its length from one end, this being a node for the vibration in question. This would, however, leave the ninth and eleventh harmonics fairly strong, their positional factors being 0·60 and 0·30 respectively. Or the ninth harmonic could be eliminated by striking the wire at a point one-ninth along its length, but then the seventh and eleventh harmonics would each sound with a positional factor 0·41. In practice, the string is usually struck at about a seventh of its distance from the end, although with a felted hammer this does not entirely eliminate the seventh harmonic; sometimes a compromise is attempted between the elimination of the seventh and ninth harmonics. But no compromise can be found which will of itself reduce both these harmonics to negligible proportions. The mainstay of the manufacturer must always be the felting on the hammer, and if this wears thin or hardens too much from prolonged use, the unwanted upper harmonics will again ring out, giving the piano a metallic or "tinny" tone. The same effect may be produced if the piano-keys are struck with undue force, so that the felt on the hammers is much compressed during its impact on the strings—indeed if we strike with absolutely immense force, every hammer, no matter how well felted, behaves like a perfectly hard hammer and produces a mere jangle of metallic tone.

The following table, calculated by Helmholtz, shews the energies of the different harmonics for a string which is struck with varying degrees of force at a point a seventh of the length of the string from one end, the force being measured by the fraction of the period of the fundamental tone for which the hammer is in contact with the string.

Relative Intensity of Harmonics (striking point one-seventh from the end)

Harmonic	String struck with a perfectly hard hammer	String struck with a soft hammer which touches string for following fraction of period of fundamental tone				String plucked
		a	b	c	d	
	0·00	0·15	0·21	0·3	0·43	
1 (c′)	100	100	100	100	100	100
2 (c″)	325	286	249	189	100	81
3 (g″)	505	357	243	108	9	56
4 (c‴)	505	260	119	17	2·3	32
5 (e‴)	325	108	26	0	1·2	13
6 (g‴)	100	19	1·3	1·5	0·01	3
7 (b♭‴)	0	0	0	0	0	0

We notice that with a soft hammer the fundamental and its octave predominate, while with a hard hammer the higher harmonics are predominant—an illustration of the general principles already explained.

For normal playing Helmholtz found that the figures given in column *d* represented the tone produced on his piano in the neighbourhood of c″, column *c* suited the region of g′, and column *b* the whole range below middle C. Thus the fundamental tone is weaker than the second harmonic throughout most of the keyboard, and is weaker even than the third harmonic through a considerable range.

The more massive hammer of the modern piano strikes the wire with greater speed than was possible at the time of Helmholtz. Because of this, the fundamental tone and the lower harmonics can be made considerably stronger than anything shewn in the above table. Indeed, the highest notes of all contain very few harmonics, the second harmonic or octave usually being the loudest component. The absence of high harmonics is no loss here, for they lie beyond the range of the human ear, and so would be inaudible even if present. The middle notes are a blend of from four to ten or more harmonics, the energy being fairly well distributed between them. The lowest notes contain very little of the fundamental tone, but are blends of many higher harmonics, as many as forty-two having been heard and identified. In extreme cases the fundamental tone may be entirely inaudible, our ears hearing only the higher harmonics, which they recombine to reproduce the fundamental in a way to be explained later (p. 241). Figs. 1 and 2 on Plate III shew two photographs taken by Professor Dayton Miller of the sound curves of the notes c″ and C of a modern piano. In the former, the fundamental note c″ is at first the loudest, and indeed almost the only, component, but after about a fifth of a second the octave appears in strength, and rapidly becomes the preponderating tone. In the latter, analysis shews that more than ten harmonics are present in appreciable strength, and continually change their relative intensities.

Much valuable information can also be obtained by analysing pianoforte tone into its constituent pure tones in the way explained on p. 59. Figs. 35, 36 and 37 shew the results of experiments by Erwin Meyer on a variety of instruments, ancient and modern.

The seven different layers of fig. 35 represent the results obtained from seven c's (CC–c′) of a modern grand

PLATE III

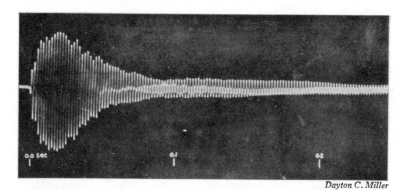

Dayton C. Miller

Fig. 1. The sound curve of the note c″ (an octave above middle C), frequency 516. The time-scale underneath shews tenths of a second, starting from the instant at which the hammer first makes contact with the wire.

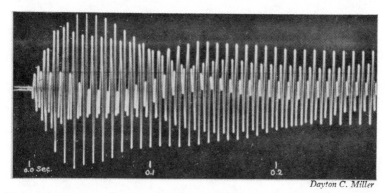

Dayton C. Miller

Fig. 2. The sound-curve of the note C (an octave below middle C), frequency 129. The time-scale is as in fig. 1.

SOUND-CURVES OF PIANOFORTE TONE

pianoforte. Frequencies are measured on a horizontal scale as marked, thus being the same for all seven layers. In each layer, the vertical line marked 1 occurs at the

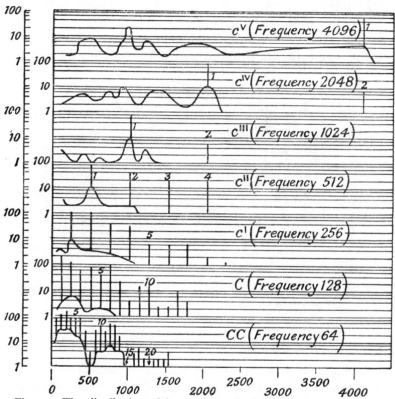

Fig. 35. The distribution of harmonics and of continuous sound in a modern grand pianoforte played *mf*.

frequency of the note sounded, while 2, 3, etc., occur at the frequencies of the second, third, etc., harmonics of these notes. Thus these vertical lines mark the positions of the various harmonics of the note sounded.

The heights of the vertical lines give the relative intensities of these various harmonics, or rather of such of them as lie within the range of the analyser used in the experiment. It must, however, be noted that the vertical height is not made proportional to the sound-intensity as measured by the physical instrument, but to the logarithm of this intensity, since this, as we shall find below (p. 224),

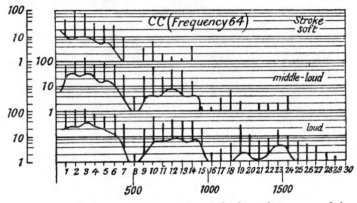

Fig. 36. The distribution of harmonics and of continuous sound in a modern grand pianoforte when played *pp*, *mf*, and *ff*.

gives a rough measure of the intensity of the sound as judged by the ear.

We must notice also that these vertical lines do not start at ground level, but from a sort of undulating mountain of sound, which represents a jumble of tones of all possible pitches. It is not altogether obvious where all this discordant and unwanted sound originates. Part of it must represent the difference between the actual piano-wire—especially the bass wire with thinner copper wire twisted round it—and the infinitely thin, infinitely flexible, string

of abstract theory. A further part can, no doubt, be traced
to the bluntness of the hammer and its coating of felt.

The distribution of sound between the various harmonics must depend (p. 92) on the strength with which the
note is sounded. The results shewn in fig. 35 were all obtained by sounding different notes *mf*. Fig. 36 shews the

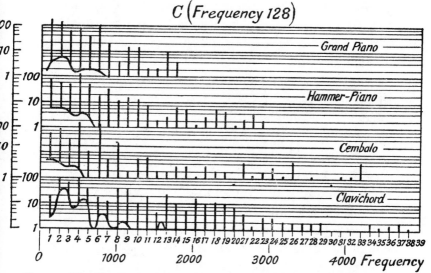

Fig. 37. The distribution of harmonics and of continuous sound in a
modern grand piano contrasted with that in various old-time instruments.

distributions obtained by playing the same note CC in three
different strengths—*pp*, *mf* and *ff*. We notice how an increase
of force on the piano-key results in more harmonics being
heard, in accordance with the principles explained on
p. 92, above.

Finally fig. 37 shews the distribution of continuous
sound and harmonics obtained when the same note, now
tenor C, is sounded on a variety of instruments. The top

layer of the diagram, which still refers to the grand pianoforte, is merely a reproduction of the sixth layer in fig. 35;
the other three layers refer to various old-time instruments.
We see that all are conspicuously richer than the modern
pianoforte in their higher harmonics, and correspondingly
defective in their lower harmonics.

Many pianists are firmly convinced that they can put a
vast amount of expression into the striking of a single note
of the piano: some claim to be able to draw the whole gamut
of emotion out of a single key. In reply, the untem-
peramental scientist points out that, in striking a single note,
the pianist has only one variable at his disposal—the
force with which he strikes the key; this determines the
velocity with which the hammer hits the wires, and once
this is settled, all the rest follows automatically. It is not,
however, a legitimate inference that single notes can differ
only in loudness; differences in the strength of striking will
also produce a difference in the proportion in which the
various harmonics enter, and this will naturally alter the
emotional quality of the note. For instance, very hard
striking increases the relative proportion of the upper
harmonics and so imparts dissonance and harshness as well
as mere loudness. It seems natural to associate the resulting
sound with anger, disappointment or despair. But it remains true that all the shades of tone which the pianist can
get out of one note form one linear sequence only, this
corresponding to the different speeds with which the
hammer can strike the wire; and it is quite certain that the
human emotions cannot be placed in a single linear
sequence. Also it seems clear that, so long as he confines
himself to striking single notes, the greatest virtuoso has no

PLATE IV

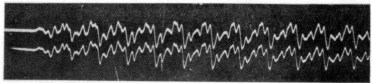

Hart, Fuller and Lusby

Fig. 1. Sound-curves of a note on the pianoforte produced: (upper curve) by a well-known pianoforte virtuoso; (lower curve) by letting a weight fall on the key. The curves are exactly similar, shewing that the virtuoso can produce no greater effect than can be produced by merely mechanical means.

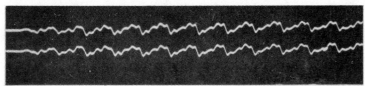

Hart, Fuller and Lusby

Fig. 2. A second pair of sound-curves produced respectively by virtuoso playing and by dropping a weight on the pianoforte key. Again the two curves are seen to be exactly similar.

SOUND-CURVES PRODUCED BY HUMAN SKILL COMPARED WITH THOSE PRODUCED BY MECHANICAL MEANS

greater range of effects at his disposal than the child strumming at its five-finger exercises.

To put this last matter beyond doubt, three American scientists, Hart, Fuller and Lusby, of the University of Pennsylvania, have recently made records of the sound curves of single notes played by well-known virtuosi, and also of the same note played by letting a weight fall on the keys. Two pairs of such curves are shewn in figs. 1 and 2 on Plate IV. In each case the upper curve records the note played by the professional, the lower curve that played by the falling weight. No visible difference can be detected.

Making the string produce the desired quality of tone is only the first problem of pianoforte manufacture; the second is the transmission of this tone to the hearer. A metal wire has so small a diameter that its vibrations transmit but little energy to the surrounding air; this is why the harp produces so feeble a sound. The pianoforte produces a fuller sound by enlisting the help of a wooden sound-board, which is made as large as the framework of the instrument permits. Securely fixed to this sound-board is a bridge over which the wires of the pianoforte pass; this transmits the vibrations of the wires to the sound-board.

When the sound-board takes up these vibrations, its large surface sets a considerable mass of air into agitation, so that the sound is heard in ample volume even at a good distance. It is important that all parts of the sound-board should be vibrating in the same phase, otherwise the vibrations from different parts of the board will neutralise one another in the way explained on p. 46. This requires that the sound-board shall be built of a wood in which sound travels very rapidly. Norway spruce, in which sound travels at three miles a second, is found to be specially suitable for this purpose. Vibrations travel the whole length of a sound-board of this material in about a two-thousandth part of a second, so that for low-pitched tones the vibrations are in approximately the same phase all over the board, and for higher tones the differences of phase are not excessive.

Bowed Strings

Theories of plucked and struck strings have proved to be comparatively simple; that of a string which is bowed, as in the violin or violoncello, is very much more intricate.

A violin-string gives out the same note when it is bowed as when it is plucked, and this shews that the bowing must set up free vibrations of the string. We could not have been sure of this from general principles alone; we might have thought that as the bow provides a continuous stream of energy, it would compel the string to perform "forced" vibrations. Such forced vibrations are, in actual fact, ruled out because the force which the bow exerts on the string has no definite periodicity of its own.

Helmholtz was able to trace the motion of a violin-string during the bowing process, by attaching a bright bead to it and taking a succession of instantaneous photographs. A more modern method is to make the bowed string perform its vibrations behind a narrow slit, placed at right angles to the string. On looking through this slit, we see only a single point of the string, and when the string is vibrating, this appears to be moving up and down behind the slit. The detailed motion is, of course, too quick to be followed with the eye, but it is easily recorded photographically. If a sensitised plate is made to move steadily and rapidly past the slit, we obtain a trace of the motion of that point of the string which lies opposite the slit.

Helmholtz found that two different types of motion alternate in rapid succession. In the first, the bow grips the string firmly, so that this is dragged along and shares the motion of the bow. This kind of motion cannot go on for ever, because the farther the string is dragged from its normal position, the greater is the force needed to keep it from slipping back. A time must come when the bow is no longer able to exert the necessary force, and then the second motion ensues. The string slips back along the bow until it has con-

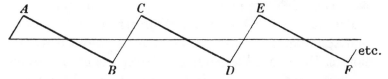

Fig. 38. Trace of the motion of a point on a vibrating violin-string, at a quarter of the string's length from the bridge.

siderably overshot its normal position, when the bow again takes hold, and the motion repeats itself indefinitely. Helmholtz found that the whole motion takes place in the plane in which the bow moves, and obtained traces of the motions of various points of the string.

Fig. 38 shews the trace for a point one-quarter of the way along the string. The parts *AB*, *CD*, *EF*, ... represent motion in which the string is being dragged with the bow; the parts *BC*, *DE*, ... that in which it is slipping back and overshooting its normal position. In this particular case the slipping movements occupy a quarter of the whole time. For other points of the string the fraction is different, being always equal to the fraction of its whole length that we have to go along the string to reach the point at which it is bowed.

Each time that the bow loses its grip on the string, as well as each time that it resumes this grip, the vibration of the string undergoes a sudden change of phase. Thus, if two violins are playing in unison, the difference in phase of their two vibrations changes repeatedly, so that the sounds they emit may reinforce one another at one instant, but enfeeble one another at the next (p. 39). These frequent alternations of loudness cause the "beating" or undulating effect which is characteristic of strings playing in unison, even when they are perfectly in tune with one another.

The data which are embodied in fig. 38 enable us to analyse the motion of the string into its constituent harmonics. Helmholtz shewed that the strengths of the various harmonics must always be in the ratio of $1 : \frac{1}{4} : \frac{1}{9} : \frac{1}{16} :$ etc., whatever the point at which the string is bowed.

This assumes that the string is bowed in a direction which is strictly at right angles to its length. Bowing in any other direction sets up longitudinal vibrations, and these make themselves heard in the agonising squeaks and scratches which characterise the efforts of the beginner, and may properly be left out of our discussion.

Violin Tone

Confining ourselves to legitimate playing, we may say that of the two factors we introduced on p. 88, the first or positional factor is always equal to unity, while the second or ordinal factor has the same value as for a plucked string. No matter where the string is bowed, the first, third, fifth and other odd-numbered harmonics occur in the same strength as in a string plucked at its middle point, but the second, fourth, sixth and other even-numbered harmonics, instead of being completely absent, are present in full strength, with the result that the bowed string has a fuller, more brilliant and richer tone than the plucked string.

It is somewhat surprising to discover that the tone quality of a violin cannot be varied by bowing the string at different points. At first this may seem to put the violin on the same level of comparative inexpressiveness as the piano. But we must remember that sounding a piano note is a momentary event—everything is settled the moment the hammer has hit the wire—whereas sounding a violin note is a continuous event. In either case the sounds can vary only in strength, but on the violin the strength may vary from instant to instant, and this opens up new dimensions of expressiveness to the violin.

Also a violin-bow has considerable width, which was disregarded in the mathematical investigation of Helmholtz. When the width of the bow is taken into account, the motion of the string is still found to be tolerably independent of the point of bowing, but is no longer entirely so. A wide bow has somewhat the same effect on violin tone as a felted hammer has on piano tone—by smoothing out the curve formed by the string, it eliminates some of the highest harmonics, particularly those which would have a node anywhere under the bow. The string may be bowed anywhere from a seventh to a fifteenth, but is usually bowed at a ninth or tenth, of its length from the bridge—more in *piano* passages, and less in *forte* passages. If it is bowed *sul ponticello*—i.e. close up to the bridge—the bow does not lie over any nodes except those of very high harmonics, so that moderately high harmonics are produced in full strength, and the note has a metallic sound. As the bow is moved farther away from the bridge, the higher harmonics sound less strongly, and the tone acquires a gentler and smoother quality which is better suited for *piano* passages. A tone of harsh metallic quality is, however, obtained by playing at any point with the wood of the bow—*col legno*—because the higher harmonics are no longer smoothed out by the width of the bow.

The body of the violin serves the same purpose as the sound-board of the piano; having picked up the vibration of the strings, it sets into vibration a larger body of air than could be affected by the strings alone. There is, however, an essential and important difference between this sound-board and that of the piano. The piano sound-board serves only to pass on the vibrations of the wires, and the more faithfully it does this the better it is deemed to be. The body of the violin, on the other hand, is expected not only to pass on the vibrations it receives from the strings, but to add something of its own. Its free vibrations are of high pitch, and as many of them coincide in frequency with harmonics of the notes produced by the strings, these particular vibrations may be much reinforced by resonance. It is their presence that gives the instrument its peculiar tone or timbre. Such a group of frequencies is known as a "formant".

Most violins have a group of free vibrations of frequencies between 3000 and 6000; and it is through the reinforcement of harmonics having these frequencies that the violin gains its distinctive rich tone. The free vibrations of the viola are of lower frequency, because of its larger dimensions, and this explains why a low note on the violin sounds quite different from the same note on the viola— the body of the violin reinforces a group of quite high harmonics, while the body of the viola reinforces a group of much lower harmonics.

It is less easy to explain why a note played on a good violin sounds different from the same note played on a bad violin. But it emerges from much discussion that, here also, a large part of the difference can be traced to a difference of frequencies in the formant. Backhaus has examined the frequencies of the body vibrations of a first-class Stradivarius, and finds that the majority are fairly evenly distributed between 3200 and 5200. In other violins the frequencies are usually lower and also less evenly distributed. A good modern violin shewed a distribution which approached that of the Stradivarius in uniformity, but the frequencies themselves were about 500 cycles lower. In a poor modern violin, the frequencies were not only less well distributed, but were also about 1000 cycles lower. In brief, the bad violin picks out certain rather low harmonics in an arbitrary way, and reinforces these unduly, while the good violin picks out a wide band of high harmonics, and reinforces these fairly impartially.

It used to be conjectured that the varnish on old violins might contribute in some special way to producing a rich even tone, but recent scientific investigations have given no support to this theory. It is the wood, rather than the varnish, which is found to be important. Lark-Horovitz and Caldwell have examined by X-rays the bodies of a number of old violins by Stradivarius, Amati and others, and find that in the best violins the front of the belly is usually made of very fibrous wood (generally spruce), cut in such a way that the vibrations do not spread as freely sideways as longitudinally, whereas the wood of which the back is made shews no such peculiarity. It seems possible that X-ray analysis may in time solve the riddle presented by these old violins, and enable manufacturers to build violins of Stradivarius quality by mass-production, if they please.

CHAPTER IV

THE VIBRATIONS OF AIR

The last two chapters have been concerned with the vibrations of tuning-forks and of strings. The vibrations of tuning-forks proved to be mainly of theoretical interest, helping us to understand vibrations and sound-curves in general. On the other hand, the discussion of the vibrations of strings established direct contact with practical musical problems, the sounds produced by the violin, piano, harp, etc. In the present chapter we shall consider a further class of musical instruments in which the vibrating structure is a column of air—organ-pipes, flutes, whistles, oboes, fifes, etc.

The Spring of Air

For our first experiment we need only very simple apparatus—an ordinary bicycle or motor-tyre pump with a reasonably close-fitting piston. Let us cork up the tube at the outlet end, and stand the pump vertically, with the piston near the top, as in fig. 39.

The piston does not immediately fall to the bottom, because the pressure of the air in the tube holds it up. We can push it down by pressing hard on the handle, but the moment we take the pressure off, it bounces up again, just as though the air inside the tube formed a spring. Indeed we have discovered what Robert Boyle called "the spring of air".

We shall understand the mechanism of this "spring of air" if we bear in mind that a gas consists of an immense number of molecules which dart about to-and-fro at very high speeds, each moving in a straight path until it either collides with another molecule or runs into some solid object. When either of these events occurs, the molecule just bounces off and starts on a new path.

The larger the molecules are, the more they will interfere with one another's motion, so that we can measure their size by examining to what extent they interfere with one another. We find that they are of very different sizes, the simplest substances having the smallest molecules, as we might perhaps expect. The simplest and smallest of all molecules, the molecule of helium, which consists of a single atom, has a diameter of rather less than a hundred-millionth part of an inch. The molecule of hydrogen, with two atoms, is rather more than a hundred-millionth of an inch in diameter, while that of air is larger still. The molecules of water vapour (H_2O) and carbon dioxide (CO_2), each of which contains three atoms, have diameters of nearly two-hundred-millionths of an inch.

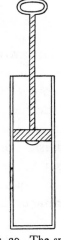

Fig. 39. The spring of air can be tested in an ordinary bicycle or car pump (closed at the outlet end). If we press the handle down it bounces up again, as though the piston were resting on a spring.

Even the smallest particle of matter we can see must obviously contain an immense number of these excessively minute molecules; in actual fact, a tiny drop of water is found to contain millions of millions of millions. If the molecules of air in an ordinary room were put end to end, they would form a chain which would go round the earth

25,000 million times; if the same molecules were spread uniformly over the surface of the earth, there would be 5000 million to every square inch of surface.

The average speed of the molecules of air of an ordinary room is about 500 yards a second, which is roughly the speed of a rifle bullet. This means that every solid surface in the room is exposed to a continuous hail of projectiles, each moving with the speed of a rifle bullet. It is this incessant bombardment that causes the pressure of the air. With each breath we take, millions of millions of molecules enter our bodies, and it is only their continual hammering on our lungs from inside that keeps our chests from collapsing. In the same way the piston in the cylinder of a locomotive undergoes millions of millions of millions of millions ($1\cdot4 \times 10^{29}$) of bombardments by molecules of steam every second; although the total weight of molecules in the cylinder is only a few ounces, yet their impact urges the piston forward in the cylinder and so propels the train of hundreds of tons weight.

We can now see what happens when we press the piston of our bicycle pump farther into the cylinder. We compress the air inside, and so crowd the molecules more closely together. As a result, the number of molecules which hit upon the piston from inside increases, and the pressure on its lower surface is increased. This increased pressure is responsible for what we have called the spring of air.

The Vibrations of a Column of Air

This quality of springiness results in a column of air having free vibrations of definite frequency—just as a metal spring has. If we blow over the open end of a pipe or tube, we hear a musical note, and the pitch of this tells us the frequency of vibration of the air inside the pipe or tube. Again, we may hold a vibrating tuning-fork over the open end of a glass vessel, while we gradually fill the vessel with water (fig. 40). At one stage of the filling process, the note of the fork may be heard to ring out clear and loud, shewing that the column of air standing above the water has a free vibration of the same frequency as the fork.

The best way of studying the free vibrations of a column of air is, however, through an experiment which is in many ways analogous to Melde's experiment. That enabled us to discover the free vibrations of a stretched string; this will enable us to discover the free vibrations of a column

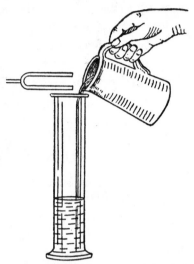

Fig. 40. When the tuning-fork gives a clear loud note there is resonance between its vibrations and those of the column of air in the glass vessel.

of air. We take a tube filled with air, closed at each end by a sliding stopper, and lay it horizontally as in fig. 41. The stopper S at one end fits closely, but the stopper T at the other is so loose that we can move it in and out easily by hand. This apparatus is practically that of the experiment with the pump, but laid in a horizontal position. We again find that we have to exert a certain amount of force to push the stopper T inwards; if we then let it go, the spring of the air inside pushes it out again, and causes it to oscillate backwards and forwards for a time, just as though there were a real spring of metal wire connecting T with S.

To find the frequencies of the free vibrations of such a column of air, we attach one fork of a vibrating tuning-fork* to the movable stopper T, and gradually move the stopper S backwards and forwards. This motion naturally changes the periods of the free vibrations of the air in the tube, and at intervals one of these must coincide with the

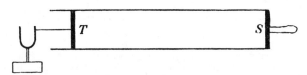

Fig. 41. Apparatus for investigation of the vibrations of a column of air ST.

period of the fork. When this occurs there is resonance, and a clear loud note rings out. We shall find that resonance occurs when the distance ST has certain definite values, and that these are all multiples of the same length. For instance, if the shortest distance for which resonance occurs is 1 foot, the other distances will be 2 feet, 3 feet, 4 feet and so on—in brief, an exact number of feet—until we reach the limit imposed by the length of our tube.

We shall see what this means if we replace our tube by one of glass (which must be very dry inside) in which we have scattered some very light fine powder, as for instance

* In the most convenient form of the experiment, known as Kundt's experiment, the tuning-fork is straightened out to form a metal rod—its longitudinal vibrations are then of far higher pitch than those of the ordinary fork, so that a comparatively short length of tube will serve.

cork filings or lycopodium powder. When the tuning-fork is set into vibration, the particles of powder exhibit the agitation of the air by dancing about, so that the whole tube is filled with dust. When the stopper reaches a position in which resonance occurs, the agitation becomes even more violent throughout most of the tube, but there are certain points of calm, at which the powder begins to settle down, and finally forms little piles. In the instance just taken, we should find that these are at distances of

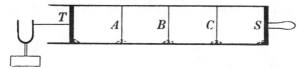

Fig. 42. If we put dust into the vibrating column of air, this settles in little piles at evenly spaced points *A*, *B*, *C* ..., shewing that these points are nodes of the vibration.

1 foot, 2 feet, 3 feet, and so on, from *T*, and as the length from *S* to *T* is an exact number of feet, the last heap of all will be close up against *S*, as in fig. 42.

It is easy to see what has happened. As the dust can stay at rest at these particular points, it is clear that there can be no agitation of the air there; these points must then be "nodes" (p. 75) of the vibration. In other words, the equal lengths of air *TA*, *AB*, *BC*, *CS* in fig. 42 must be vibrating separately—just as the stretched string vibrated in separate equal lengths (p. 67).* Thus the free vibrations

---

* It may seem paradoxical that *T* is a node, although the loose stopper is moving backwards and forwards, keeping the air in motion there. The explanation is that the stopper has a far greater weight than all the air in the tube, so that a quite small motion of the stopper imparts enough energy to the air to keep it in violent agitation. Thus the amplitude of the stopper is small compared with that at other parts of the tube, and *T* is virtually a node, or very nearly so.

of a column of air are exactly analogous to those of a stretched string; the column can vibrate in any number of equal lengths.

We might equally well have performed the experiment by fixing the stopper S, thus keeping the column of air of constant length, and picking out the frequencies of its free vibrations by substituting a succession of tuning-forks at the other end. We should then have found that the frequencies consist of a fundamental frequency and harmonics having frequencies 2, 3, 4, ... times that of the fundamental. The vibration having four times the frequency of the fundamental is, of course, one in which the column of air vibrates as four separate equal parts.

From a series of experiments of this kind, we shall find that the period of each vibration is exactly proportional to the length of the column of air which is vibrating. This is the exact analogue of the law of Pythagoras (p. 64) and admits of a similar interpretation. For, just as with the stretched string (p. 69), we can regard the vibrations of a column of air as made up of waves travelling through the air from T to S, and back again after reflection at S. The law then tells us that these waves travel always with the same speed.

We shall understand this better if we again think of the air in the tube as a collection of swiftly moving molecules. Let us suppose that when the tuning-fork is first set into motion, it begins by pushing the stopper T farther into the tube. This compresses the gas in the layer TT' (fig. 43) immediately behind T, and so not only increases the intensity of bombardment on T, but also the bombardment on the next layer $T'T''$. So far the bombardment across T'

has been an equal battle—as many molecules have crossed from right to left as from left to right—but it now becomes unequal. More molecules begin to cross T' from the left than from the right. This relieves the congestion in TT', but only at the expense of creating a congestion in the next layer $T'T''$. Exactly the same thing now happens in this layer as had previously happened in the layer TT', so that the congestion is passed on to a third layer $T''T'''$, and so on indefinitely.

Thus the inward motion of the stopper T causes a wave of congestion to travel along the tube. It must obviously

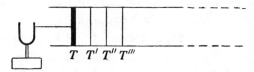

Fig. 43. The passage of a wave through a column of air.

take time to travel, because the molecules themselves only travel at a finite speed. When T first begins to move to the right, molecules are pushed away from T and, so to speak, carry the news of what has happened to the farther layers —rather like an army in flight. This news cannot possibly travel faster than the messengers which carry it, namely the molecules, and actually it does not travel quite so fast, because the molecules do not dash forward in uninterrupted straight paths, but are continually buffeted about by the other molecules which collide with them, and so pursue a zigzag course. After allowing for this and various other considerations, we find that the wave of congestion travels at only 74 per cent. of the average speed of the molecules.

After the stopper *T* has moved a certain distance to the right in fig. 43, the tuning-fork reverses its motion, and the stopper begins to move to the left. The motion just described now repeats itself, except that the wave is no longer

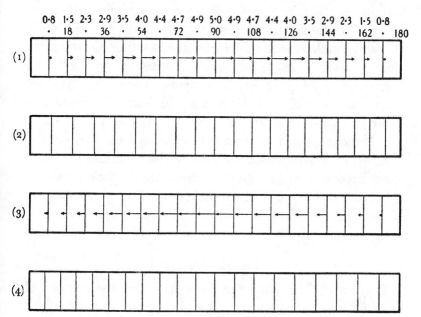

Fig. 44. Four stages in the vibration of a column of air. The vertical lines represent the same particles in different stages of the motion. The arrows shew the direction of motion, their length being proportional to the speed of motion. (The figures in the top line are proportional to the lengths of the arrows below them; the second line of figures represents the difference in phase between the point and the extreme left-hand.)

one of compression, but of rarefaction. It travels at the same speed as the wave of compression which preceded it, namely at 0·74 times the average speed of the molecules.

The repeated backwards and forwards motion of the tuning-fork sends a succession of waves of compression and rarefaction through the gas, all travelling at precisely the same speed. Each wave, as it reaches the stopper, is reflected back, and travels along the tube in the reverse direction. The superposition of two waves of equal amplitude travelling in opposite directions constitutes one of the free vibrations we have already discussed—the motion is exactly analogous to the vibration of a string illustrated in fig. 30 (p. 71). Fig. 44 shews the distribution of density at four stages of progress. The arrows shew the speed of motion; the closeness of the lines indicates the density of the gas, and the order is 1, 2, 3, 4, 1, 2, 3, 4, 1, ..., etc.

If the stopper S were not blocking up the far end of the tube, the waves would pass out of the tube into the open air beyond, and the sound of the vibrating fork would be heard through the whole of the surrounding space. Clearly then this travelling of waves of alternate compression and rarefaction must constitute the passage of a wave of sound.

The Speed of Sound

It is easy to calculate the speed at which these waves of sound travel through the tube. In the simplest vibrations of all, the wave travels from T to S and back again while the fork makes half a complete vibration, so that it travels four times the length of the pipe in the course of a whole vibration. In this way we find that the speed of the wave is about 1100 feet a second.

This is well known as the speed with which sound travels in air. We may often see a puff of steam coming from the whistle of a distant train, but it is not until a few seconds later that the sound of the whistle reaches us. On comparing the distance of the train with the time its sound has taken to reach us, we shall find that the speed of sound is about 1100 feet a second, or twelve miles a minute. Or, if we happen to live about two miles from Westminster, we may listen to the chimes of Big Ben on our radio, and then hear the sound of the actual bells come through our open windows some ten seconds later. Again, we can calculate that sound travels about twelve miles a minute, or 720 miles an hour.

Although this speed is rapid in comparison with most of the speeds we meet in life, it is not unthinkably rapid; indeed, it is only about double the speed of the fastest aeroplane or motor-car. We can obtain a visual demonstration of it by watching troops marching behind a military band. As each man hears the beats of the music, he puts his foot down, and so slightly lowers his head, but the men do not all hear the beats at the same time, because the sound of the band takes time to travel along the line of troops. We see a wave of head-lowering running along the line of men as the sound reaches them, just as we see the heads of corn lowered in a wheatfield when a gust of wind passes over them. Just as the motion of the wave of head-lowering in the corn shews the speed of the wind, so the motion of the wave of head-lowering in the troops shews us the speed of sound.

The fact that sound travels only at a finite speed introduces certain complications into the performance of music. In a large orchestra, two instruments will often be as much as 50 feet apart, and in a large, and especially in a divided organ, two manuals may have their pipes 50 feet apart. As sound takes about a twentieth of a second to travel 50 feet, two sounds may be produced simultaneously and yet a listener may hear one a twentieth of a second later than the other.

Now a twentieth of a second is not a negligible quantity in the performance of music; at ♩ = 152, it is the duration of a ♬, and a time-lag of this amount may reduce a trill or a rapid passage to an unintelligible blur of discordant sounds. For this reason instruments which will sound together in an orchestra should be placed in as close a physical proximity as is possible, and the organist should think well before coupling two manuals of which the pipes stand far apart.

So long as a sound is not too loud, its speed of travel is the same for all intensities. Very violent noises, such as explosions and gunfire, are found to travel considerably faster than the quieter sounds of music, but we are not concerned with these in the present book. Also the speed of travel is the same for musical sounds of all pitches—when music is played at a distance, the different notes reach us precisely in the order in which they are played; a chord remains a chord, and does not spread out into an arpeggio, as it would if there were any tendency for notes of high or low pitch to scramble in front or lag behind the notes of medium pitch.

These two facts are of great importance to music. If there were any tendency for loud notes or soft notes, high notes or low notes, to travel faster or slower than others, all music would be reduced to chaos before it reached the listener.

The Speed of Sound in Gases other than Air

If we repeat the experiment described on p. 111, but fill our tube with some gas other than air, such as domestic coal gas, we shall find a different speed of sound, and the same is true if we fill the tube with warm air. The following table shews the results of experiments of various kinds on the speed of sound in different gases:

Gas	Speed of sound
Dry air at 32° F.	1087 feet a second
,, 60° F.	1118 ,,
,, 212° F.	1287 ,,
Hydrogen at 60° F.	4340 ,,
Carbon dioxide at 60° F.	850 ,,

If we again think of our gas as a collection of swiftly moving molecules, we see at once why sound must necessarily travel at different speeds in different gases—it is because their individual molecules travel at different speeds. A general law of physics tells us that heavy molecules move, on the average, more slowly than light ones, so that sound travels more slowly in a gas with heavy molecules than in one with light molecules, as is shewn in the above table. The table also shews that sound travels faster through a hot gas than through a cold one, and we can now see why. To warm a gas, we must supply extra energy to it. This extra energy distributes itself over the various molecules of the gas, causing each to move at a higher speed, and so increasing the speed at which sound is transmitted through the gas. In ordinary air, the speed of molecular motion increases by approximately one per cent.

for every ten degrees Fahrenheit that the temperature rises, so that the speed of sound increases by 1·1 feet a second for each degree.

The fact that the speed of sound varies with the temperature entails important practical consequences. We have seen that the period of vibration of a column of air is proportional to the time sound takes to travel over the length of the column. If the air is warmed, sound travels faster and the period becomes less. It follows that the pitch of all wind instruments is raised when the temperature rises, or when they are taken into a warmer atmosphere. This explains why the instruments of an orchestra must be tuned afresh each time the temperature changes. Before a concert the tuning is performed in the concert hall itself, so that the various instruments will be in tune with one another in the actual air in which they are to be played, and, for the same reason, the players of wind instruments breathe into their instruments before tuning them.

Finally, the fact that sound travels at different speeds in different gases provides a means of discovering when the air of a coal-mine is vitiated by "fire-damp"—the explosive gas which may cause a disaster if it comes into contact with a naked light. Two similar pipes or whistles are blown simultaneously in the mine, the one being filled with pure air which has been brought into the mine in a metal container to serve as a standard, the other with the air of the mine whose purity is under suspicion. If the air of the mine contains much fire-damp, the sound will travel at different speeds in the two pipes, so that the free vibrations of the two columns of air will be of slightly different frequencies, and beats will be heard. The number of beats heard per second gives, of course, a measure of the degree of impurity of the air.

Refraction of Sound

It is a general property of wave-motion that waves travelling through a uniform substance follow a straight path; when the substance is not uniform, they are bent or "refracted" somewhat as a ray of light is bent in passing from air to water. The bending is always away from the substance in which the speed of travel is fastest, just as, when troops wheel round, the direction of march is bent away from the man who walks fastest. Suppose, then, that we have a layer of cold air near the ground, with a layer of hot air lying above it. The speed of travel is faster in the upper layer, so that when sound which has been travelling through the lower layer strikes the upper layer, it is bent back towards the lower layer, and may be driven back into this and compelled to continue its journey through it.

These conditions often occur over the surface of a lake or other piece of still water, especially in the early morning when the water of the lake is still cold. If a person in a boat speaks or sings, the sound of his voice will begin by spreading out in all directions, but as soon as the waves reach the upper layer, they are bent back and forced to continue their journey through the lower layer. No energy is dissipated by an upward spreading of the waves, so that the voice can be heard to a far greater distance than it otherwise could.

The same conditions, but on a larger scale, often prevail in hilly or mountainous country, so that a voice can be heard to incredible distances, the sound not so much crossing the valleys as creeping along the ground.

On a larger scale yet, the same conditions are found in the atmosphere as a whole. This consists of a lower layer known as the troposphere, and a higher layer known as the stratosphere. The latter is much warmer than the former, so that when a sound is generated down below, only a small fraction of its energy passes into the stratosphere, the remainder being bent back and coming to earth again. An explosion or other loud noise may often be heard at great distances by waves of sound which have been reflected back from the warm stratosphere.

Somewhat similar conditions may also occur when the wind is stronger high up above the earth's surface than it is near to the ground. Suppose that a wind from the west is blowing at 20 feet a second near the ground, and at 40 feet a second higher up. Then a sound which is produced near the ground will travel through the air at 1100 feet a second, while the air is itself travelling eastward at 20 feet a second. Thus sound will travel eastward at 1120 feet a second when near the ground, but at 1140 feet a second higher up. Again the sound is continually bent away from the layer of faster travel—the upper layer—and so is compelled to creep along the ground, and may be heard to great distances. For sound travelling westward, exactly the opposite conditions prevail. The speed of travel is 1080 feet a second near the ground, and 1060 feet a second up above. The sound is continually bent away from the earth's surface, and may become quite inaudible at only a short distance to the west of its origin. We see why sound so often seems to be "carried with the wind", and why it is difficult to "shout into the wind".

Air Vibrations in Music

A column of air is like a stretched string in having vibrations of which the frequencies stand in the simple ratio $1 : 2 : 3 : 4 : \ldots$ In either case the harmonics are the "natural harmonics", so that anything which causes the fundamental tone to sound is likely to produce a generous supply of harmonics as well.

This property explains the outstanding importance of the stretched string and the column of air as sources of musical sound; if they are kept continuously in vibration, their higher harmonics, being natural harmonics, are sounded as well as their fundamental tones, and we hear the rich musical tone which results from an abundance of concordant harmonics.

We have also seen that other structures do not possess this property; the frequencies of the free vibrations of drums, cymbals and triangles do not stand in any simple ratio to one another, so that their harmonics are not natural harmonics. For this reason it would be difficult to make their higher tones sound continuously, and even if they did, they would produce discordant tones. In brief, their sounds are only suited for momentary hearing. We see at once why the instruments of an orchestra fall into the three departments of strings, wind and percussion, and it is interesting to notice that these correspond exactly to the bow-string, the broken reed and the drums and beaten sticks which, it has been suggested, first awakened the musical feelings of primitive man. We begin to see why all the musical instruments of to-day have developed out of these three prototypes—why thousands of years of effort have been unable to find new departments of music-producing instruments.

The special property of the free vibrations coinciding with the natural harmonics is possessed to perfection by the air inside a tube which is closed at both ends. Unhappily the vibrations of such a column of air can only be excited by opening the tube to the outer air at one or both ends. The perfection then disappears, for a reason we shall discuss shortly. If the diameter of the tube is very small in comparison with its length, the perfection is only slightly impaired, and the vibrations are still not unlike those of a stretched string—this is why the string-toned registers of an organ are made of excessively narrow pipes. In the more usual case, in which the diameter of the pipe is appreciable, there is a deficiency of the higher harmonics—this is why wide organ-pipes of the diapason type need to have their harmonics supplemented by separate smaller pipes which supply the missing harmonics directly (p. 244).

We can only discuss such problems in detail, when we understand the way in which the vibrations of a column of air are excited. We accordingly turn to a discussion of this question.

Whirlpools and Whirlwinds

When water is flowing in a rocky torrent, we may notice a great difference in the quality of flow before the stream encounters a rock and after. Before it meets the rock, the torrent flows onwards in a calm steady stream; afterwards it is broken into innumerable whirlpools and eddies. We see the same thing when a boat or ship moves through still water; in front of the bow there is calm water, but astern

PLATE V

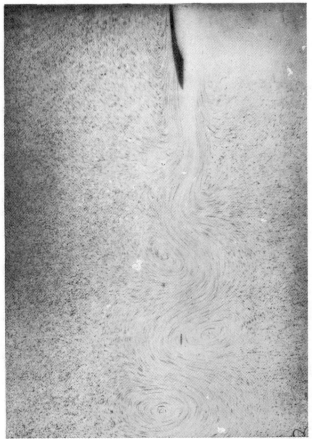

G. J. Richards

EDDIES FORMED BY DRAWING AN OBSTACLE
THROUGH STILL WATER

The motion of the water is made visible by mixing into it
minute drops of milk and alcohol.

PLATE VI

G. J. Richards

THE SAME EDDIES AS ARE SHEWN IN PLATE V, BUT AT
A GREATER DISTANCE BEHIND THE OBSTACLE

The procession of eddies is now perfectly regular, and it is clear that eddies
have been formed on the two sides of the obstacle in turn.

there is a seething mass of whirlpools. The general features are much the same as those we see when the water flows past a fixed rock; indeed they must be, since it can make but little difference whether the water moves past an object at rest, or the object moves through water at rest.

Plates V and VI shew photographs by G. J. Richards of the eddies formed behind an obstacle when it is drawn through still water. The eddies immediately behind the

Fig. 45. The eddies formed in water as it streams past a long thin wire.

obstacle are shewn on Plate V, those farther behind on Plate VI. Fig. 45 shews a sequence of six drawings of eddies formed when water streams past a circular wire. Both this and the photographs produced on Plates V and VI shew that the eddies are formed on the two sides of the wire alternately.

Exactly similar phenomena occur in air, although they are not so easily observed. When the wind or a blast of air encounters a small obstacle, little whirlwinds are formed which are the exact counterparts of the whirlpools which are formed when a stream of water strikes a rock. There is a steady flow of air in front of the obstacle, and a steady train of whirlwinds behind it. These whirlwinds are formed on the two sides of the obstacle alternately; as soon as one comes into existence, it begins to drift away in the general current of air, thus making place for others which are formed in turn behind it. Some drawings of such whirlwinds are shewn in Plate VII (facing p. 133).

We may seem to be still a long way from music. Actually we are very near, for it is precisely these little whirlwinds of air that are responsible for the production of sounds in wind instruments—without them our flutes and organpipes would cease to function.

The "Wind Whistle"

When whirlwinds are formed by the wind streaming past an obstacle of any kind, the formation of each little whirlwind gives a slight shock, both to the obstacle and to the air in its neighbourhood. If the wind blows in a continuous steady stream, these shocks are given to the air at perfectly regular intervals. We may then hear a musical note—it is what is often described as the "whistling of the wind", or the "wind whistle". Its pitch is of course determined by the frequency of the shocks to the air, and this is the number of whirlwinds formed per second. Experiment shews that a whirlwind is formed every time the wind passes over a distance equal to $5\frac{2}{5}$ times the diameter of the obstacle, and this makes it possible to calculate the pitch of the note. Suppose, for instance, that we are at sea, with the wind blowing at 40 miles an hour through a rigging of half-inch ropes. Simple arithmetic shews that 40 miles an hour is 704 inches a second, so that the wind traverses 1408 diameters of the rope every second. Dividing this by $5\frac{2}{5}$, we obtain 261 as the frequency of the note of the "wind whistle"—middle C of the piano. If the wind blows faster, whirlpools are formed faster and the pitch of the wind whistle rises, the frequency being exactly proportional to the wind velocity. When the wind "howls", we hear the pitch of the note rising and falling, and its frequency at any instant gives a measure of the speed of the wind at that instant. If the obstacles which the wind meets are smaller, the pitch is higher; this is why we hear notes of high pitch when the wind blows over the telegraph wires on land, and still higher notes when it blows through stalks of corn or blades of grass.

Each little whirlwind gives a shock not only to the air, but also to the obstacle to which it owes its existence, so that, as the whirlpools are formed on alternate sides of it, this is pushed to-and-fro from side to side. It is these pushes that make the rope of a flagstaff flap in the breeze, while the fluttering of the flag at the top shews the whirlwinds chasing one another along it, first on one side and then on the other.

These motions are all "forced" vibrations, being forced by the flow of the wind. The solid obstacle will, of course, have its free vibrations as well, and if the period of any one of these happens to coincide with the period of the vibrations forced by the wind, resonance will occur, and a musical note may ring out very loud and clear. This is why the telephone wires sing at their loudest in frosty weather; the cold has contracted them, so that they are stretched as tight as violin-strings, and their free vibrations have gone up into the region of frequencies inhabited by the whirlwinds.

The Aeolian Harp

It is these same little whirlwinds that produce the sound of the Aeolian harp. In this a number of wires are stretched across a framework, which may be placed in an open window or in any place where there is likely to be a good draught of air. As this blows across the strings, it sets them into vibration in the way just explained, and a clear note is heard whenever one of the vibrations set up by the wind has the same frequency as one of the free vibrations of the strings. The strings are usually tuned to the same pitch, but are made of different thicknesses, and so emit wind whistles of different pitches when the wind blows over them. A string sounds clearly and fully when the pitch of the wind whistle it emits coincides with any one of the harmonics of the note to which the strings are tuned. So long as the wind confines itself to exciting harmonics not higher than the tenth, we hear sounds which belong to the ordinary musical scale. The eleventh and higher harmonics, however, introduce notes which do not belong to the ordinary scale, and these provide the weird unearthly quality we associate with Aeolian tones.

Edge Tones

Tones of a similar nature are produced when a stream of air or gas strikes the sharp edge of a wedge of metal or other hard substance. The phenomenon has been studied in great detail by Lootens, Hensen, Weerth, Wachsmuth and many others. The general procedure is to maintain the air in a reservoir R (fig. 46) at a steady pressure, and allow a thin blast of air to escape out of a narrow slit S and strike upon a sharp edge E placed parallel to it.

The impact of the stream of air on this sharp edge produces practically the same physical conditions as occur when the wind blows on a telegraph wire, or on a string of an Aeolian harp. Little whirlwinds are formed on the two sides of the edge alternately, and move along the sides of the wedge to make place for their successors. Again the whirlwinds are formed at perfectly regular intervals, and so produce a musical note of definite pitch, known as an "edge tone". We have seen that the frequency of an

Fig. 46. Apparatus for the study of edge tones.

Aeolian note is proportional to the speed of the wind. In the present instance the speed of the air blast replaces that of the wind, and this clearly diminishes as we go farther from the slit. Thus the pitch of an edge tone will depend on the distance of the edge from the slit. If the jet began to spread fanwise the moment it emerged from the slit, we should expect the velocity at E to be inversely proportional to the distance SE, in which case the frequency of the note sounded would vary inversely as the distance SE. Wachsmuth has verified that this is approximately the case, but only up to a certain limit. If the edge E is gradually moved away from the slit, the note falls in pitch for a time in accordance with the law just explained. But when the distance SE reaches a certain critical value, the pitch, instead of continuing to fall, suddenly jumps an octave.

A detailed study of the eddies shews that each has now lost its former simplicity and broken into two, single eddies no longer being long enough to span the gulf. The eddies are now formed just twice as frequently as before, so that the frequency of the edge tone is doubled, and its pitch rises by an octave.

After this the law of inverse distance is again obeyed for a time, and then the note again makes a jump—this time through the interval of a fifth, and so to the twelfth, or third harmonic, of the note our simple law would lead us to expect. Clearly each of the original eddies has now broken into three. Later on, yet another jump occurs, this time to the fifteenth or super-octave of the note given by the law of inverse distance. Thus as the distance SE is gradually increased, the "edge tone" sounds in turn the successive harmonics of the fundamental tone given by this simple law. The reason for the successive jumps of pitch is always the tendency for the simple whirlwinds to break up into a number of smaller units.

We can pick out these different edge tones by using a set of Helmholtz resonators. Or we may place a tube near to the edge E, and if one of the free vibrations of the column of air in this tube happens to have the same frequency as the edge tone, this tone will sound clearly and firmly from the tube.

Flue Organ-Pipes

A note produced in this way is practically identical with the note of an ordinary flue organ-pipe. In this the slit, the edge, and the resonating column of air are all combined to form a single structure. The arrangement is shewn diagrammatically in fig. 47.

Air is maintained at a steady pressure in a wind-chest *W*, and can only pass from the wind-chest into the pipe when a valve or "pallet" *V* is opened. A blast of air then passes through the foot of the pipe *F*, and is concentrated by a slit *S*—known as the "flue"—into a narrow jet which impinges on the upper lip *L*.

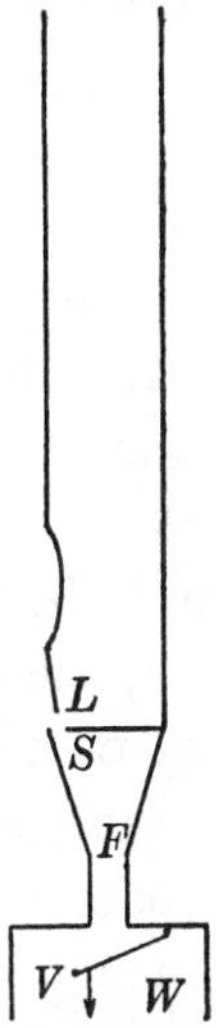

Here eddies are formed in the way already explained, and cause rapid alternations of pressure, which produce "edge tones" and also set up vibrations in the column of air in the tube. If the edge tone is approximately in resonance with one of the free vibrations of this column of air, energy will be supplied to this particular vibration very rapidly, so that the pipe will speak promptly. If the periods are widely different, energy will accumulate slowly in the free vibration, and the pipe may be slow of speech.

Usually there is no close approximation to resonance between the edge tones produced at the lip of the pipe and the body vibrations of the column of air in the pipe. The two form what is known as a "coupled system"—a system formed of two separate systems, neither of which can perform its own free vibrations without interference from the other.

A system is said to be "loosely coupled" when this

Fig. 47. A flue organ-pipe. Air escaping through the slit *S* impinges on the lip *L* and produces an edge tone, thus setting the column of air inside the pipe into vibration.

mutual interference is slight, so that either system can perform its own free vibrations almost unmolested by the motions going on in the other. The free vibrations of the coupled system are then the sum of the separate free vibrations of the two component systems, except that the pitches of either may be pulled just a little out of tune by the presence of the other. A system is said to be "closely coupled" when the opposite conditions prevail.

If one of the two systems is far more massive than the other, its vibrations will usually have much more energy. The more massive system will then "force" its vibrations on the weaker, so that the free vibrations of the whole system differ but little from those of the more forcible partner, each being pulled a little out of tune by the weaker partner.

A simple illustration of a coupled system is provided by two clocks which are placed in such close contact that each influences the time-keeping of the other. If they are placed at opposite ends of a mantelpiece, they provide an instance of loose coupling; their mutual influence may be so slight that each keeps its own time, almost—although never quite—uninfluenced by any vagaries there may be in the time-keeping of the other. Bringing them nearer increases the closeness of their coupling; if we finally place them back to back this may become so strong that they keep exactly the same time. If they are of equal strength and size, the time they keep will be exactly half-way between the two times they would have kept if they had been going independently. If, however, one is much stronger and more massive than the other, it will take control of the whole motion. Its pendulum sets up vibrations in its own frame-

PLATE VII

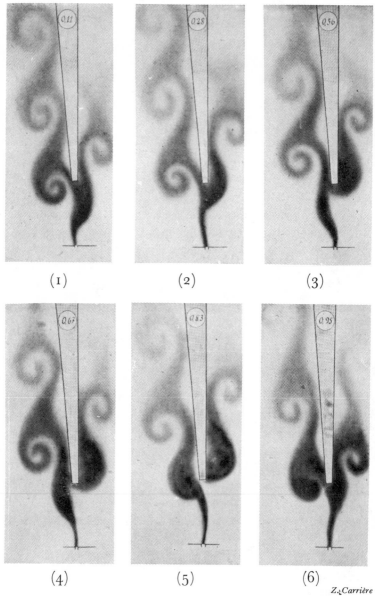

(1)   (2)   (3)

(4)   (5)   (6)

Z.Carrière

EDDIES FORMED AT THE LIP OF A FLUE ORGAN-PIPE

The number on each drawing gives the phase as a fraction of the whole period. The motion of the air was made visible by mixing fine smoke with it.

work, and these are transmitted to the framework of the weaker clock, with the result that the oscillations of the pendulum of the smaller clock are no longer "free", but are "forced" to keep pace with the vibrations of the framework from which it is suspended. In this way the weaker clock is forced to follow in the footsteps of its stronger brother, and both will keep the time which the stronger clock would keep if the weaker one were non-existent.

The edge tones and the vibrations of the column of air in a flue-pipe form a coupled system of this latter kind. The energy of the latter vibrations is so much greater than that of the edge tones, that the latter may almost be disregarded, and the vibrations of the whole structure treated simply as those of the air in the pipe. Nevertheless, the edge tones are just too strong to be disregarded entirely, and exert a certain slight influence on the tone of the pipe as a whole. We have seen that blowing the pipe with a stronger blast of wind must raise the pitch of the edge tone, and this is found to affect the pitch of the note emitted by the pipe to a perceptible extent. In flutes, fifes and other wind instruments, the edge tones and air vibrations are rather more closely coupled, so that the expert performer can control the pitch within fairly wide limits by skilful blowing.

Plate VII shews some drawings, made by Carrière, of the trains of whirlwinds at the mouth of an organ-pipe. The pipe lies to the right in each drawing, and the six pictures are taken at approximately equal intervals of a sixth of a period, so that in the repeating cycle

$$1, 2, 3, 4, 5, 6, 1, 2, \ldots$$

all the intervals are approximately equal. The resemblance

to the photographs already shewn on Plates V and VI (pp. 124, 125) shews that edge tones are produced by substantially the same mechanism as Aeolian tones.

Carrière has also made some important observations on the way in which the pitch of the edge tone varies with the height of the mouth of the organ-pipe—the interval SE in fig. 46 or SL in fig. 47. The table below shews the results of a set of observations in which the slit was $\frac{1}{10}$ inch (2·5 mm.) in width and the pressure of the air was that of a column of water 160 mm. (6·3 inches) high.

Height of mouth	Frequency of edge tones	Product
81 mm.	80	6480
89	74	6586
99 (= 4 inches)	62	6138
109	54	5886
119	48	5712
129	43	5547
139	35	4865
149 (= 6 inches)	33	4917
159	30	4770
167	28	4676

We see that the frequency falls off as the height of the mouth increases, and so far this is in accordance with the law already mentioned. On the other hand, the two changes are not strictly proportional, as is shewn by the product, given in the last column, not being exactly constant. We get a much better proportionality if we subtract 40 mm. from the height of the mouth; we then find that the product of the reduced height of mouth and the frequency is approximately constant—as though the jet of air travelled for about 40 mm. before beginning to spread out fanwise and lose its speed.

Carrière has also studied the effect of changes in wind pressure. The table below shews the results he obtained when the slit was kept at the constant distance of 119 mm. and the pressure was made to vary from 30 to 200 mm. water.

Pressure (mm. of water)	Frequency of edge tone
30 = (1⅛ inches)	23
40	25
50 = (2 inches)	28
60	30
80	33
100	35
120	40
140	42
158	44
182	46
200 = (8 inches)	48

We see how an increase in the pressure, which necessarily increases the speed of the air blast, increases the frequency of the edge tone also.

As the vibrations of the column of air in an organ-pipe involve much more energy than the edge tones, they almost pull the latter into their own frequencies. Nevertheless, for the pipe to sound promptly and with a clear full tone, it is desirable that there should be as good resonance as possible between the edge tone and the vibration tone of the column of air in the pipe. To attain this demands the skill of the "voicer". For instance, if the frequency of our pipe is 33 (CCC), the tables just given shew that we can get an edge tone in perfect resonance with it from either of the following combinations:

Pressure = 80 mm. water; height of mouth = 119 mm.

or Pressure = 160 mm. water; height of mouth = 149 mm.,

besides which there must of course be an infinite number of other combinations not entered in the above tables. A low wind pressure naturally demands a low mouth, so that the speed of the air shall not have fallen too much before reaching the mouth, and vice versa.

The two above tables are for a slit $\frac{1}{10}$ inch in width. If the width of the slit is increased, the pitch of the edge tone rises somewhat rapidly, because a broad jet of air retains its velocity longer than a narrow one, and so strikes the lip with higher speed. Thus a lowering of pressure can be compensated by increasing the width of the slit.

When the pallet under a pipe is opened, the blast of air does not rise to its full speed or full pressure at once. Thus a pipe which has been adjusted to give complete resonance when the pressure has attained its full value will not usually be in good resonance while the pressure is building up to this value. Indeed, during this interval we frequently hear edge tones which are quite different from the natural note of the pipe—sometimes the octave or twelfth, and sometimes an even shriller chirping sound, which seems to consist of high harmonics of the lower edge tone produced by the lower pressure. Some of the older continental organ-builders appear to have looked on the presence of this tone with favour, possibly because they did not know how to eliminate it. The modern voicer usually tries to suppress it, sometimes by filing a number of small notches or nickings in the tongue of the pipe. These increase the effective width of the slit at the points at which they occur, and so give a stronger air blast at these points. In some cases these may compensate for the defect pressure, so that the exact edge-tone needed to set up resonant vibrations is present almost from the outset, and the speech may be so prompt that the discordant edge tones are not heard at all. In other cases the phenomenon cannot be so simply explained.

Stopped and Open Pipes

Flue-pipes may be either "stopped" or "open". In a stopped pipe, the end remote from the mouth is closed by a tight-fitting stopper. There can be no motion of the air here, so that this end of the pipe is necessarily a node, while the mouth, at which there can be no variations of pressure,

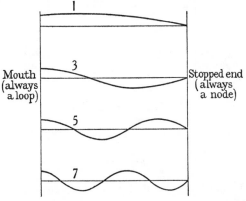

Fig. 48. The modes of vibration of a stopped organ-pipe. The displacement at any point is represented by the distance between the thick curve and the horizontal line nearest it.

is a loop. Thus the possible modes of free vibration for the air inside a stopped pipe are those shewn in the above scheme (fig. 48).

For the first mode of vibration, the wave-length is four times the length of the pipe. This gives the fundamental note of the pipe. For the second mode, the wave-length is only a third of that of the fundamental mode, so that the frequency is three times that of the fundamental, and the note sounded is the third harmonic of the fundamental. The other modes of vibration sound the fifth, the seventh, the ninth harmonic, and so on. Thus a stopped pipe emits odd-numbered harmonics only, like a string plucked at its middle point.

If we now remove the stopper from the pipe, we shall find that the pipe emits the even-numbered harmonics only, the odd-numbered harmonics disappearing as soon

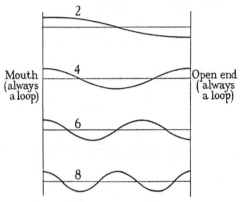

Fig. 49. The modes of vibration of an open organ-pipe.

as the stopper is withdrawn. The reason for this is that the open end of the pipe is no longer a node, but has become a loop because the pressure is that of the outside air. The scheme of vibration is now that shewn in fig. 49.

If we forget that our pipe ever had the stopper in it, we shall think of the top vibration (marked 2) as the fundamental note of the open pipe, although it is of course the octave of the fundamental note of the stopped pipe. The other modes of vibration (marked 4, 6, 8, etc.) now appear as the second, third, fourth, etc., harmonics of this new fundamental note, so that the pipe is sounding all the harmonics of its own fundamental note.

This, however, is only approximately true. In the first place, the open end of a pipe is not exactly a loop; we must go a short distance beyond the open end to become entirely clear of the constraining influence of the pipe and find a pressure precisely equal to that of the free air. Thus the loop lies a short distance beyond the actual end of the pipe, and the wave-length of the note sounded is slightly more than double the length of the pipe. This small addition to the effective length of a pipe is known as the "open-end correction"; for a cylindrical pipe its value is found by experiment to be approximately 0·29 times the diameter of the pipe.

In the second place, it is even less legitimate to treat the mouth of a pipe as a loop, and a still larger "mouth correction" must be made here. According to Cavaillé-Coll, the total correction to be made amounts to about 1⅔ diameters for a cylindrical pipe, whether open or stopped, and to about double the internal depth for a square pipe, whether open or stopped. Thus an open pipe 8 feet in length and 6 inches in diameter (diapason) gives out a fundamental note of wave-length 17 feet 8 inches, while a stopped wooden 4-foot pipe of 6 inches depth (bourdon) gives a fundamental note of wave-length 20 feet.

Just as in the experiment described on p. 112, the column of air in such a pipe is capable of vibration in any number of separate parts, but the open-end and mouth corrections introduce a complication of some consequence. This can perhaps be best explained by a numerical example. Suppose that the air inside a pipe has a fundamental vibration of frequency 100. When this same air vibrates in two separate parts, these vibrations will not each have a frequency of 200, as they would if there were no open-end and mouth corrections. For the corrections are not quite the same for the quicker vibrations as for the fundamental tone, so that instead of having frequencies of 200, these may have frequencies of, let us say, 202. Thus the pipe as a whole may have frequencies of perhaps 100, 202, 305, and so on. The closeness of the coupling will usually draw the edge tone into agreement with the fundamental tone of the pipe, so that this will exert a force on the air of the pipe which repeats itself 100 times a second. Fourier's theorem now shews that this may be regarded as made up of a number of forces which repeat in simple harmonic fashion at the rate of 100, 200, 300, ... times a second respectively. Thus what we may describe as the higher harmonics of the edge tone will not be in perfect resonance with the higher harmonics of the air in the pipe, with the result that these upper harmonics may only sound feebly.

As this killing of the higher harmonics is a consequence of the open-end and mouth corrections of a pipe, it is only pronounced when these corrections are large in amount. In pipes of very small diameter, the corrections are so small as to be insignificant, and the higher harmonics are heard in abundance. For this reason, the string-toned stops of the organ—the viol d'orchestre, viola da gamba, etc.—in which high harmonics must figure prominently, are made of pipes of very small diameter. On the other hand, the flute-toned stops from which high harmonics must be excluded are usually made of pipes of large diameter. There are, however, other ways of excluding the higher harmonics. The old German "Spitz-flöte" was formed of pipes which were of small diameter, but tapered very considerably at the top, this giving an open-end correction which was very large, because conditions were a good step on towards those of a stopped pipe. The Boehm flute as usually played in the orchestra employs the same artifice, its tube being constricted internally near the mouth. The tones of the highest register of all on this instrument are found on analysis to be almost pure tones without harmonics, as might be expected since the diameter of the vibrating column of air is comparable with its length.

The proportion in which the different harmonics occur in the note emitted by any pipe, open or closed, depends on other factors also, especially the wind pressure and the height and shape of the mouth of the pipe. By increasing the wind pressure or lowering the mouth of a pipe, or even by sharpening its upper lip and so lowering its resistance to the air-blast, the organ-builder can increase the speed with which the air jet strikes the lip of the pipe, and so raise the pitch of the edge tone. This procedure naturally increases the strength of the upper harmonics; if it is carried to extremes the fundamental note may disappear altogether, so that an open pipe will speak its octave, and a stopped pipe its third harmonic, the twelfth. In general, mouths which are cut low give notes which are rich in all the harmonics if the pipes are open, and in all the odd-numbered harmonics if the pipes are stopped. Registers of stopped pipes are found, especially on continental organs (Quintaten, Quintadena, etc.), in which an effort is made to obtain a large proportion of the third harmonic (twelfth). Similarly, registers of open pipes are found in which the octave is developed to a high degree. Open "harmonic" pipes have a small hole bored at a point half-way along the pipe, so that this point becomes a loop, and the pipe sounds mainly its octave, the fundamental being heard only as a faint growl. There are also harmonic stopped pipes (Zauberflöte, etc.), in which a hole is bored at a little more than half (usually about nine-sixteenths) of the length of the pipe above the mouth. Again the fundamental tone almost disappears, but in this case the pipe speaks its twelfth.

On the other hand, a pipe with high mouth and blunt upper lip can be made to give a tone from which the higher harmonics are almost absent—pure flute tone.

Reed Organ-Pipes

The pipes so far discussed produce their sound in the same way as the flute and piccolo of the orchestra, namely through a column of air being excited into vibration by an edge tone. The organ also contains other pipes in which the production of sound is like that in the clarinet or oboe of the orchestra, the column of air inside the pipe being set into motion by the vibration of a reed.

The reed is rather like the reed of an ordinary harmonium. This consists of a spring of metal which is screwed down tightly at one end A (fig. 50), and is shaped to fit closely into an aperture in a rigid piece of metal, which lies between a lower wind-chest W, and an upper wind-chest C. When the appropriate stop of the harmonium is drawn, air under pressure fills the wind-chest W and spreads round the reed into the upper wind-chest C. In the top of this latter chest is a second opening, which is normally covered by a felted

block of wood D. When the appropriate key of the harmonium is depressed, the block D is raised and air under
pressure escapes from C. As the reed now has a greater
pressure of air below than above, it is forced upwards, air
rushing past it from the lower wind-
chest W to the upper wind-chest C.
But before the pressures in the two
chests have become fully equalised,
the elasticity of the reed carries it
back to its original position, so that
the flow of air is checked and the
pressure in W again increases. By a
continued repetition of this process, the reed is set into
violent vibration, with a period equal to that of its free
vibration.

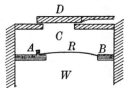

Fig. 50. The reed of the
harmonium.

The reed of an organ-pipe differs from this mainly in
being just too large to fit into the metal aperture, and so lies
against it, striking it at intervals. For this reason it is called
a "striking" reed, while the harmonium reed is described
as a "free" reed. Also the valve which sets the reed in
action is not above the reed as in fig. 50, but is inside the
wind-chest W.

This arrangement leaves the farther side of the reed free,
and a pipe is placed here of such length and shape that the
period of its free vibration coincides approximately with
that of the reed. When the reed is set into vibration, the air
in the pipe, being in resonance with it, is also set into
vibration, and speaks with a clear firm note.

It is this resonating pipe which is responsible for the
essential difference between the tone of the harmonium
and that of the organ. The reeds of both instruments have

one principal free vibration, and a number of others which, as is usual in solid structures of metal, are discordant with this. When the harmonium reed speaks, all these tones are heard together and produce a discordant noise. In the organ, the main body of sound does not come from the reed, but from the air in the pipe. This is in resonance only with the main vibration of the reed, so that it reinforces this particular vibration alone, and what we hear is this vibration combined with the harmonics of the pipe. If the pipe is skilfully designed, these will form a concord.

Ellis has made a mathematical investigation of the proportion in which the different harmonics will be heard. He finds that in cylindrical pipes only the odd-numbered harmonics are reinforced by resonance; this explains why the pipes of the clarinet stop, as also the tubes of the orchestral instrument, are made cylindrical. A conical pipe on the other hand makes no such discrimination, odd and even harmonics being equally reinforced; for this reason the pipes of the oboe, trumpet, tromba, etc., and the tubes of the instruments they imitate, are

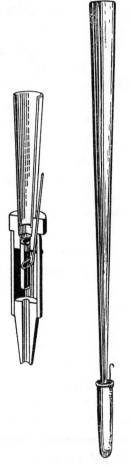

Fig. 51. A reed-pipe of an organ. The complete pipe is shewn on the right, a section of the lower part containing the reed on the left.

made conical in shape. If a brilliant tone is desired, the reed motion itself must supply plenty of higher harmonics to be reinforced by resonance; this is why the organ-builder uses a "striking" reed in preference to a free or harmonium reed, the sudden stoppage of the air jet providing exactly the high harmonics needed. In old organs this stoppage was absolutely sudden, so that the upper partials were exceedingly prominent, producing a fiery, but also very harsh, tone. The modern organ-builder prefers to curve his reed at the end, so that it does not "strike" and cut off the air jet quite suddenly, but rather unrolls itself and cuts off the air jet gradually. This makes the upper partials less prominent, and the tone smoother and less fiery.

Early writers on the action of flue-pipes pictured the edge tone as being produced by the vibrations of a "reed of air". Although this view must now be discarded, it is obvious that the formation of eddies of air on alternate sides of the lip have much the same effect as the in-and-out motion of a reed. In the flue-pipe, these eddies and the air in the pipe form a coupled system. In the reed-pipe, the reed and the column of air form a coupled system. The column of air is still the predominant partner, but as its predominance is not so marked as in the flue-pipe, a change in the frequency of vibration of the reed has a marked effect on the pitch of the note emitted by the pipe as a whole. So much is this the case that reed-pipes are usually tuned by altering the frequency of the free vibrations of the reed.

We have already seen that any change of temperature alters the speed of sound in air very appreciably—by about 1·1 feet a second, or a thousandth of the whole, for each degree Fahrenheit of change. It has, however, far less effect on either the dimensions or the elasticity of wood or metal. Thus when the temperature changes, we may almost disregard any change produced in the dimensions of pipes or resonators, or in the size and elasticity of reeds. The only thing that changes appreciably is the column of air in each pipe, and the frequency of its vibrations will change in exactly the same ratio as the speed of sound. For instance, a rise of 30 degrees Fahrenheit will increase the frequency by 3 per cent. of the whole, and this sharpens the pitch by half a semitone. The flue-pipes of the organ all have their pitches changed to approximately the same extent, and so stay fairly well in tune with one another; actually small pipes are slightly more affected than large by changes of temperature.

The resonators of the reed-pipes also change in exactly the same way, but as the reeds themselves remain practically unaffected by the change of temperature, they pull down the frequency of the complete pipe, with the result that the instrument as a whole sounds out of tune.

If there is a uniform rise of temperature throughout the instrument, the flue-pipes will stay in tune with one another, and as they form the main body of the instrument the performer may think that the instrument has stayed in tune except for the reeds which appear suddenly to have gone flat.

If, however, flue-pipes stand in parts of the organ which are at different temperatures, even these will go out of tune with one another. This is a reason for leaving the swell-boxes open after playing, so that the whole organ shall be at a uniform temperature for the next performance, since air imprisoned in a closed swell-box cannot follow the temperature changes of the outer air.

Orchestral Wind Instruments

Practically all wind instruments are similar, in the general method of their operation, to one or other of the two classes of organ-pipes we have just described, so that it is hardly necessary to discuss the workings of each instrument in detail. There is, however, one general question that must be touched upon.

Organ-builders usually specify the precise nature of the metal or wood of which their pipes are to be built, the reason being that the quality of tone depends on the material of the pipe. For instance, pipes of wood produce a heavier, but also a warmer and more mellow, tone than pipes of metal, while pipes of nearly pure tin produce a richer tone than pipes of cheaper metal. The same is even more true of orchestral instruments; a silver clarinet sounds very different from one of wood, just as an orchestral flute sounds different from a penny whistle.

If the sound were produced merely by the vibration of a column of air, such differences as these could not arise; the air would vibrate in the same way no matter what material was used to enclose it. The fact that differences of timbre can be heard shews that the pipe must itself contribute something to the production of the sound. The pipe has of course its own free vibrations, their frequencies depending naturally on the material of which it is made. Clearly, then, some of these must be reinforced by resonance with the vibrations of the column of air.

Sound is known to travel much faster in solids than in air; we have, for instance:

Velocity of sound in air		= 1,100 feet a second
,,	lead	= 4,100 ,,
,,	tin	= 8,300 ,,
,,	oak	= 14,000 ,,
,,	Norway spruce	= 16,000 ,,

Thus a pipe of wood or metal will in general have vibrations of much higher frequencies than the column of air it contains, and if the vibrations of the pipe are in resonance with any of the harmonics of the air, these latter must be quite high harmonics. Thus the free vibrations of the pipe itself constitute a sort of formant similar to the formant of the violin described on p. 104.

In the wind instruments of the orchestra, the same pipe is used to produce notes of very different pitch, but the formant, which depends only on the structure of the pipe itself, remains always the same. Clearly the formant has much to do with the characteristic timbre of the instrument; some writers even claim that the timbre of the instrument is completely dominated by it.

Fig. 52, which has been compiled from experiments by Hermann-Goldap, shews the formants heard when certain instruments are played. The range of fundamental notes is shewn by the thick line to the left, that of the formant by the dotted line on the right. The numbers over the latter lines give the intensity of the notes of the formant in terms of that of the fundamental note; this increases as we pass down the table, being smallest for the clarinet and flute and greatest for the trumpet and oboe.

The presence of these higher harmonics is clearly seen

in the wave-forms of the instruments in question. The photographs shewn on Plates VIII and IX were all taken by Professor Dayton Miller, and are reproduced by his courtesy. They shew the sound-curves of notes of the flute, the oboe, the saxophone and the clarinet, played under various conditions.

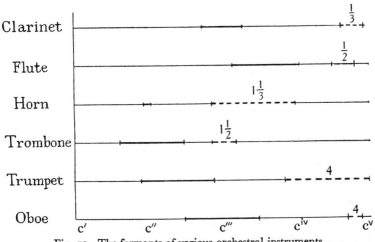

Fig. 52. The formants of various orchestral instruments.

Fig. 1 on Plate VIII shews the sound-curves of the note b′ played *p*, *mf* and *f* on a flute. The topmost curve, obtained by blowing the instrument softly, is almost a pure simple harmonic curve, shewing that the fundamental tone predominates. The middle curve obtained by playing *mf* is of less regular shape, shewing that harmonics are superposed on to the fundamental tone, when the instrument is blown louder. Finally the bottom curve, produced by playing *f*, contains still more harmonics; exact analysis shews that the second and third harmonics predominate.

Fig. 2 on the same plate shews the sound-curve of middle C played on an oboe, and fig. 1 on Plate IX shews the sound-curve of G♯ played on a saxophone. In both these curves, the deep and close indentations of the curve indicate the admixture of high harmonics in great strength.

In fig. 2 on this plate the upper curve is the sound-curve of middle C of frequency 257, played *mf* on a clarinet. It is much smoother than the two preceding curves of the oboe and saxophone, shewing that the higher harmonics are not present in any great strength. But clarinet tone is not always as smooth as this. The lower curve in the same figure was produced by playing the same note on a different clarinet. We see sequences of very deep indentations, which occur intermittently, one series occurring for each wave of the fundamental tone. This is exactly what we should expect from the beats of near or consecutive harmonics. Careful measurement shews that $11\frac{1}{2}$ of these short depressions occupy the same length of curve—i.e. of time—as a single wave of the fundamental tone, so that we may tentatively identify the beating harmonics as the eleventh and twelfth (approximately $f\sharp^{iv}$ and g^{iv}). Professor Miller has verified this identification by analysing the sound-curve into its various harmonic constituents. He finds that the relative amplitudes of the different components are as follows:

First harmonic	c'	$=29$	Second harmonic	c''	$= 7$
Third ,,	g''	$=20$	Fourth ,,	c'''	$= 1$
Fifth ,,	e'''	$= 2$	Sixth ,,	g'''	$= 6$
Seventh ,,	$b\flat'''$	$= 6$	Eighth ,,	c^{iv}	$= 8$
Ninth ,,	d^{v}	$=16$	Tenth ,,	e^{iv}	$= 9$
Eleventh ,,	$f\sharp^{iv}$	$=30$	Twelfth ,,	g^{iv}	$=35$

It is of interest to notice that the eleventh and twelfth

PLATE VIII

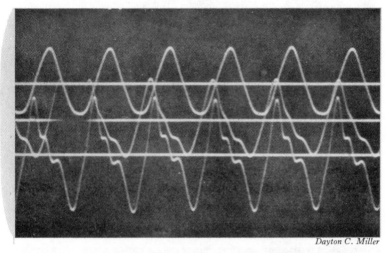

Dayton C. Miller

Fig. 1. Sound-curves of a flute played *p* (top), *mf* (middle) and
f (bottom). The note is b′ of frequency 488.

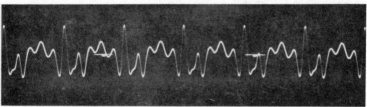

Dayton C. Miller

Fig. 2. The sound-curve of an oboe, played *mf*. The note is middle C of frequency
254. The two time-scale marks half way up the photograph are $\frac{1}{100}$ second apart.

SOUND-CURVES OF FLUTES AND OBOES

PLATE IX

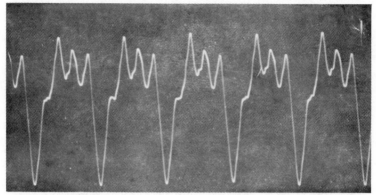

Dayton C. Miller

Fig. 1. The sound-curve of a saxophone. The note is G♯ of frequency 209.

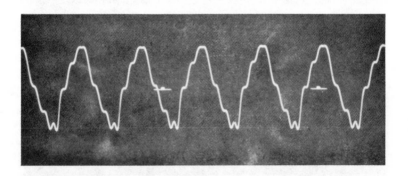

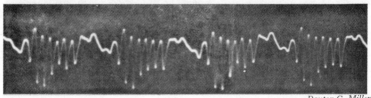

Dayton C. Miller

Fig. 2. The sound-curves of middle C played on two different clarinets.

SOUND-CURVES OF SAXOPHONES AND CLARINETS

harmonics, of frequencies 2827 and 3084, are both stronger than the fundamental note of frequency 257, and as the ear is many times more sensitive to notes of the higher frequencies than to notes of the lower, the sound which the ear perceives must consist almost entirely of tones of these higher frequencies. We see from fig. 52 that these higher tones are close up to the range of the formant.

CHAPTER V

HARMONY AND DISCORD

In the present book, we are dealing with subjects which lie partly within the province of science and partly within that of art, and the boundary between the two provinces is not always perfectly clear. If the question is debated as to whether the music of John Sebastian Bach is superior to that of his son Philipp Emanuel, science can bring nothing to the discussion. The question is purely one for artists, and it is quite conceivable, although perhaps rather improbable, that they may not be able to agree as to the answer. On the other hand, if the question is whether the music of either Bach is superior to that produced by a chorus of cats singing on the roof, there will be little doubt as to the answer. The artists will all agree, and science is able to explain to a large extent why they agree.

To say the same thing in another way, the aim of music is to weave the elementary sounds we have been discussing into combinations and sequences which give pleasure to the brain through the ear. As between two pieces of music both of which give pleasure in a high degree, only the artist can decide which gives more, but the scientist can explain why some give no pleasure at all. He cannot explain why we find Bach specially pleasurable, but he can explain why we find the cat music specially painful. And this brings us to the subject of the present chapter— why is it that some combinations of sounds are agreeable to the ear, while others are disagreeable?

Through Beats to Discord

On p. 49 we imagined two tuning-forks sounded together. The pitch of one was kept fixed at 261 vibrations to the second, while the pitch of the other started at 262, and was gradually raised. As the pitch rose beats were heard, for a time, but subsequently could no longer be distinguished as such. The sound of the combined tones began by being pleasant to the ear, but as the number of beats per second increased, it gradually became more unpleasant. The unpleasantness reached a maximum when there were about 23 beats to the second, and then began to decline. Brues, extending the experiment into the region in which beats can no longer be distinguished, finds that the decline is only slight, and that, broadly speaking, the unpleasantness remains at a fairly uniform level until the octave of frequency 522 is reached, at which point it suddenly disappears.

If the same experiment is performed with violin-strings, very different results are obtained. The unpleasantness no longer stays at a fairly uniform level, but fluctuates wildly. It almost vanishes at the interval of a major third, and again at the interval of a fourth, while it disappears completely at the intervals of the fifth and octave. At the exact points at which these minima of unpleasantness are reached, the frequency ratios of the variable to the fixed tone are found to have the simple values $5:4$, $4:3$, $3:2$ and $2:1$.

Concord associated with Small Numbers

It is found to be a quite general law that two tones sound well together when the ratio of their frequencies can be expressed by the use of small numbers, and the smaller the numbers the better is the consonance. This will be clear from the following table, in which the intervals are arranged in order of increasing dissonance:

Interval	Frequency ratio	Largest number occurring in ratio
Unison	1 : 1	1
Octave	2 : 1	2
Fifth	3 : 2	3
Fourth	4 : 3	4
Major Third	5 : 4	5
Major Sixth	5 : 3	5
Minor Third	6 : 5	6
Minor Sixth	8 : 5	8
Second	9 : 8	9

and so on.

In brief, the farther we go from small numbers, the farther we go into the realms of discord. This was known to Pythagoras 2500 years ago; he was the first, so far as we know, to ask the question, "Why is consonance associated with the ratios of small numbers?" And although many attempts have been made to answer it, the question is not fully answered yet.

The central Pythagorean doctrine that "all nature consists of harmony arising out of number" provided of course the simplest of all answers, but only by building on an unproved metaphysical basis. An answer on equally uncertain foundations was given by the Chinese philosophers of the time of Confucius, who regarded the small numbers 1, 2, 3, 4 as the source of all perfection.

Euler's Theory of Harmony

In 1738 the mathematician Euler attempted an explanation on psychological lines, saying that the human mind delights in law and order, and so takes pleasure in discovering it in nature. The smaller the numbers required to express the ratio of two frequencies, the easier it is—such was his argument—to discover this law and order, and so the pleasanter it is to hear the sounds in question. Euler went so far as to propose a definite quantitative measure of the dissonance of a chord. His plan was to express the frequency ratio of the chord in question by the smallest numbers possible, and then to find the smallest number into which all these could be divided exactly. This last number, he thought, gave a measure of the dissonance of the chord. For example, the frequency ratio of the notes of the common chord C E G c′ is 4:5:6:8. The measure of dissonance is accordingly 120, since this is the smallest number of which 4, 5, 6 and 8 are all factors.

It is easy to criticise this theory from all sides. In the first place it fails to explain the facts, since it assigns the same measure of dissonance, namely 120, to the chord of the seventh C E G B (with frequency ratios 8 : 10 : 12 : 15) as to the far less dissonant common chord. Again if we put one note, say E, out of tune by one per cent. of its frequency (about a sixth of a semitone) we increase Euler's measure of dissonance 100-fold; if we now reduce the out-of-tuneness to a tenth of this, we *increase* the measure of dissonance another tenfold. If one note is only infinitesimally out of tune, the measure of dissonance at once shoots up to infinity, which is a complete *reductio ad absurdum*. Finally, Euler's theory fails to explain why we enjoy hearing the common chord, with its 120 units of annoyance, when we could reduce the annoyance to 24 units by dropping E out of the chord, and could eliminate the annoyance altogether by sitting in silence. It must be admitted, however, that this is a defect of most theories of discord. Innumerable theories are ready to tell us the origin of the annoyance we feel on hearing a discord, but none even attempts to tell us the origin of the pleasure we feel on hearing harmony; indeed, ridiculous though it may seem, this latter remains one of the unsolved problems of music.

If we were compelled to attempt a solution, it would perhaps be somewhat on the following lines. The exercise of any of his faculties gives pleasure to a healthy being—otherwise he would never attempt crossword puzzles or mountain ascents—and the greater the use made of the faculty the greater the pleasure, at any rate within limits. We like to hear CG rather than C because the irritation produced by the very slight discordance of the notes is far less than the pleasure added by the hearing of the G. On the other hand, we do not enjoy hearing CC♯, because the annoyance is so great that the balance swings in the opposite direction.

D'Alembert's Theory of Harmony

So far all theories of harmony had been either arithmetical or metaphysical. The first attempt at a physical theory of harmony originated with another mathematician, d'Alembert

(1762), who admitted his indebtedness to some earlier specu-
lations of Rameau (1721). Their theory was based on the fact
that every fundamental tone heard in nature is accompanied
by its second harmonic (the octave), by its third harmonic
(the twelfth), and so on. The interval between the octave
and twelfth being a fifth, they argued that it was "most
consonant to the scheme of nature" that two notes a fifth
apart should sound together, and so on.

Helmholtz's Theory of Harmony

Then Helmholtz (1862) developed a theory of consonance
and dissonance in terms of beats—a theory which has been
much discussed and criticised, but still holds the field
to-day. We have already seen that C and C♯ sound badly
together because they make unpleasant beats. In the case
of wider intervals such as C and F♯ there are no beats to be
heard, either pleasant or unpleasant, but Helmholtz asserted
that C and F♯ sound badly together because certain of
their harmonics (e.g. g′ and f′♯) make unpleasant beats.
On the other hand C and G sound well together because
few of their harmonics beat badly:

$$C \quad c' \quad g' \quad c'' \quad e'' \quad g'' \text{ etc.}$$
$$G \quad g' \quad d'' \quad g'' \quad b'' \quad d''' \text{ etc.}$$

indeed many harmonics are common to both notes. On
this theory the octave becomes the most perfect of all
concords, since none of the harmonics can possibly beat
worse than when one note is sounded alone. The theory is
sometimes stated in the slightly different form that two
notes sound well together when, and because, they have
certain harmonics in common, but this form of statement

overlooks the annoyance which may be introduced by such harmonics as are not possessed in common.

The theory explains at once why the dissonances of tuning-forks (p. 153) are so completely different from those of musical instruments—the tuning-forks have no upper harmonics to make beats with one another.

A few simple experiments with orchestral instruments, or at the keyboard of the organ, will convince us of the essential soundness of the theory. If we draw a flute-stop, and sound the chord C E G c', we hear no perceptible dissonance. If we now sound the same chord on a stop in which the harmonics are more developed than in the flute, the dissonance is more marked; the dissonance must have been introduced by the harmonics, since nothing else has been added which could have introduced it. It is noticeable on the diapason, and becomes unpleasant on the trumpet or clarinet. Finally it becomes intolerable on the mixture, a stop which consists of harmonics and nothing but harmonics; we shall, for instance, hear c'' (the second harmonic of c') beating badly with b' (the third harmonic of E), and g'' (the third harmonic of c') with g♯'' (the fifth harmonic of E). If we hold the chord C E G c', and add suitably chosen stops in succession, we shall hear the dissonance growing *pari passu* with the harmonic development.

It is the same in the orchestra; chords which sound well on the flutes or strings are impossible on oboes and clarinets. We understand the reason for this as soon as we notice the rich harmonics shewn in Plates VIII and IX.

Helmholtz attempted to test his theory by calculating the amount of dissonance it implied for different intervals.

He first assumed a law, somewhat arbitrarily, for the amount of dissonance produced by the beats of two pure tones at an assigned distance apart, and was then able to calculate, by simple addition, the total dissonance produced by all the beats of all the harmonics of a pair of notes. This naturally depended on the proportions in which the different harmonics entered into each tone; Helmholtz assumed the proportion to be that of violin tone.

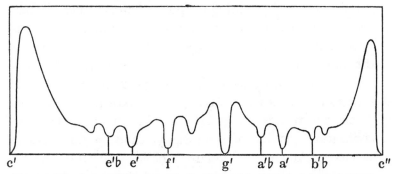

Fig. 53. The degree of dissonance, as calculated by Helmholtz, of two violin tones sounding together. The lower tone c′ sounds continuously, while the upper tone moves gradually from c′ to c″.

The result of his calculation is shewn in fig. 53. One violin-string is supposed to sound c′ continuously, while the pitch of the second ranges from c′ to c″. The degree of dissonance at any point is shewn by the height of the curve above the horizontal line in fig. 53. Obviously the main consonances and dissonances within the octave are reproduced with remarkable fidelity.

A still better test can be made by employing the exact experimental results of Brues to give the dissonance of pure tones, but the final result is much the same as that just given.

The Origin of the Musical Scale

With all this in our minds, let us try to imagine how the different musical scales may have come into being. No one seems to know precisely how music itself came into human life, but it probably was through either stringed instruments or wind instruments. Primitive man may have enjoyed the rhythm he could make by pounding sticks together, or even by beating primitive drums, which he may have used for marking time for dances or marches, but it seems likely, as already suggested, that he first discovered the pleasures of tonal music by hearing vibrating strings— perhaps the twang of his bowstring—or sounding pipes, such as the wind whistling over the top of a broken reed. Ancient drawings and reliefs shew him, in the infancy of civilisation, playing both on the lyre and on pan-pipes or syrynxes. At Ur Sir Leonard Woolley unearthed the remains of an eleven-stringed lyre, which proves that 5000 years ago or more, man had already passed from the enjoyment of a single musical sound to that of a succession of sounds. Two pictures of bands of Sumerian musicians of 4600 years ago are reproduced as the frontispiece of this book, and explain themselves. An Egyptian painting of about 2750 B.C. shews a complete orchestra of seven players, two of whom are playing on stringed instruments and three on wind instruments, while two in the middle seem to be engaged in clapping their hands as though to beat time— the discovery that the right number of conductors to an orchestra was one had yet to come, but mankind of 5000 years ago was at least acquainted with melody. He may even have been acquainted with harmony as well, although this is far from certain. For in the most primitive civilisation of all, music seems always to have been homophonic (one-part), as it still is to some extent among the Chinese, Indians, Turks, Arabs and even Greeks of to-day.

In time, however, the idea must have occurred to sing or play two or more notes at once—possibly because it was impossible for men and boys to sing together in the same pitch, or possibly because one-part music began to pall. Up to now, the exact pitches of the notes selected to form the scale had been almost a matter of indifference; from now on it was important that two or more notes of the scale could be sounded together without undue dissonance. Even to-day, many of the races which have not advanced beyond homophonic music—as for instance the Arabs, Persians and Javanese—use scales whose notes are not at all consonant; the dissonance is harmless because two notes are never heard together. On the other hand, even primitive races whose music is polyphonic use scales in which most intervals are consonant.

The octave, the simplest and most perfect consonance of all, must have been discovered at a very early stage; it is fundamental in the music of all peoples, even the most rudimentary. The early Greeks seem to have employed no other concord in their music, although they were certainly acquainted with others. Aristotle tells us that the voices of men and boys formed an octave in singing, and asks "Why is only the consonance of the octave sung, for this alone is played on the lyre?" He suggests that other consonances were not in favour because "both tones are concealed, one by the other", and compares part-music to many speakers "who are saying the same thing at the same time, when we should understand a single speaker better", which seems to suggest that he did not possess a true polyphonic ear.

Nevertheless, the time was bound to come when incessant movement in octaves was found sterile and uninteresting—witness the scholastic prohibition of consecutive octaves, which was subsequently extended to consecutive fifths also. We can imagine our primitive musician discovering the consonance of the fifth c–g, possibly first introduced, as some have thought, in the form of a descending fourth, c'–g or f–c. This, with the octave, would give a set of four strings c f g c', of which any two except f g could be played together without creating unpleasantness. Nicomaeus tells us that down to the time of Orpheus, lyres were tuned to sound these notes.

The possessor of such a lyre would still have no great variety of tones for melody, so that we can imagine him increasing the number of his strings, and planning that each new string should create no new unpleasantness, or at any rate as little as possible, when sounded together with the already existing strings. On a lyre sounding c f g c', two of the strings c, c' can be sounded with either of two new tones f and g without creating discord; the other two strings f and g can be sounded with only one, since c and c' introduce the same new tone. Our pioneer might try to give another possibility of concord to g by introducing its fifth d. The problem then repeats itself, and to find a new and pleasant companion to d, he introduces a. So he goes on and gets a succession of notes F-C-G-D-A-E-B etc., each of which can be sounded in perfect harmony with either the note preceding it or the note succeeding it in the sequence. But there is always the trouble that the first and last note have only one agreeable companion, and so we can imagine him pressing on until finally, when he has gone far enough, he finds his sequence repeating itself. His notes no longer form a straight-line sequence but a circle of notes, which can be arranged like the numbers on the face of a clock, as in fig. 54, so that each string has now two notes with which it can sound in perfect harmony, the one in front and the one behind.

It need hardly be said that the foregoing account has been purely fictitious; if for no other reason, because there was no single primitive man, but vast numbers of tribes and peoples who developed music independently, and in the most varied surroundings. But all were

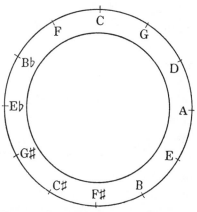

Fig. 54. The clock-face of twelve notes— the twelve semitones of the octave. Each note sounds the harmony of a fifth when played with either of the notes next to it.

striving for the same goal, and the principles which guided them—to choose pleasant noises rather than unpleasant, consonances rather than dissonances—must have been precisely those which we have imagined guiding our fictitious primitive man, so that they were led to much the same result as he, and this with a unanimity which is remarkable. They exhibit enormous differences in their language, customs, clothes, modes of life and so forth, but all who have advanced beyond homophonic music have, if not precisely the same musical scale, at least scales which are all built on the same principle.

The main differences are found in the numbers of notes which form the scale. By stopping at different places in the sequence F-C-G-D-A-...,* we obtain the various scales which have figured in the musics of practically all those races which have advanced beyond the one-part music of primitive man.

The first three notes of the sequence C, F and G formed the main tones of the scale of ancient Greece. If we proceed as far as five notes C, D, F, G, A we have the pentatonic scale in which a considerable amount of Chinese and ancient Scottish music is written, as well as much of the music of primitive peoples in Southern Asia, East Africa and elsewhere; transpose it a semitone up, and we have the scale provided by the black keys of the piano—hence the fact, beloved of school-children, that many Scottish melodies, "Auld Lang Syne", etc., can be played without touching the white keys at all, and that almost any sequence of notes strummed on the black keys sounds like a Scottish melody. On taking the first seven notes, we have the ordinary diatonic scale, which seems to have been introduced into Greece in the middle of the sixth century B.C., was standardised by Pythagoras, and has remained the normal scale for western music ever since. The beginnings of this scale cannot be traced. Garstang found two Egyptian flutes, the date of which cannot be later than about 2000 B.C.; these gave the seven-note

---

* There is no theoretical reason for starting with F rather than with any other note of the scale. F has been selected merely in order to keep off the black notes of the piano for as long as possible.

scale C D E F♯ G A B, which is identical with the Synto-lydian scale of ancient Greece.

Finally, the full clock-face of twelve notes supplies the complete chromatic scale of modern music.

The Problem of Temperament

This last scale is affected by a serious complication. The tones which give the best concord after the octave have their frequencies in the exact ratio 3 : 2, or 1·5, as is evident from their constituting the second and third harmonics of the same fundamental note. Thus, for perfect concord, each step of one "hour" on the clock-face would increase the frequency by a factor 1·5, and twelve such steps would increase it by a factor of $(1·5)^{12}$, of which the value is 129·75. We have just said that these twelve steps bring us back to the c seven octaves above the c from which we started. But we now see that they do not bring us back exactly; the frequency of this last c is only 128 times that of our starting-point, so that our twelve steps slightly overshoot the mark, and bring us to a note whose frequency is greater than that of c by a factor of 1·0136; this interval is commonly known as the "comma of Pythagoras", and is rather less than a quarter of a semitone.

To put the same thing in another way, we have just identified the frequency ratio 1·5 with the interval of a fifth, although our table (p. 25) gave the value as 1·4983. The difference is only small—1·13 parts in a thousand—but by the time we have taken the twelve steps needed to pass completely round the clock-face, it has been multiplied twelvefold into the difference of 13·6 parts in a

thousand, which represents the aforesaid difference in pitch
of almost a quarter of a semitone. When this is allowed for,

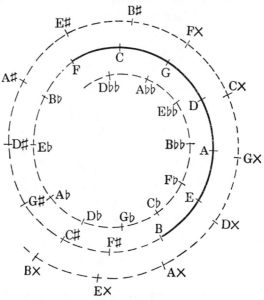

Fig. 55. The form assumed by the clock-face of fig. 54, when true fifths are
used. There is now an endless series of notes, such notes as B♯, C and D♭♭
being of different pitch.

the true clock-face is that shewn in fig. 55; it extends to
infinity in both directions, and all simplicity has dis-
appeared.

The Pythagorean Scale

Various ways have been suggested for avoiding this complication. When Pythagoras standardised the musical
scale in mathematical terms, he did not encounter it at all,
because he did not think of his series of notes as forming
a closed circle. He assigned to his C exactly 1½ times the

frequency * of F, to his G exactly $1\frac{1}{2}$ times the frequency of C, and so on, thus arriving at a scale with the ratios shewn in the following table:

Pythagorean frequency ratio	Pythagorean interval	Equal temperament frequency ratio (p. 177)
C $\quad= 1\cdot0000$		$1\cdot000$
	Tone	
D $\quad\frac{9}{8}= 1\cdot1250$		$1\cdot1225$
	Tone	
E $\quad\frac{81}{64}= 1\cdot2656$		$1\cdot2599$
	Hemitone	
F $\quad\frac{4}{3}= 1\cdot3333$		$1\cdot3348$
	Tone	
G $\quad\frac{3}{2}= 1\cdot5000$		$1\cdot4983$
	Tone	
A $\quad\frac{27}{16}= 1\cdot6875$		$1\cdot6818$
	Tone	
B $\quad\frac{243}{128}= 1\cdot8984$		$1\cdot8877$
	Hemitone	
C $\quad= 2\cdot0000$		$2\cdot0000$

From the way in which the scale is formed, it follows, as a matter of pure arithmetic, that the intervals C D, D E, F G, G A and A B must all be exactly equal, with a frequency ratio of 9 : 8. Pythagoras described each of these intervals as a "tone", and was left with the two smaller intervals E F and B C, each of which is represented by the more complicated frequency ratio of 256 : 243, or $1\cdot0535$. Pythagoras called such an interval a "hemitone". It is distinctly less than either the half of a Pythagorean tone or the modern semitone, the frequency ratios being

$$\begin{aligned}
\text{Pythagorean hemitone} &= 1\cdot0535, \\
\text{Half of Pythagorean tone} &= 1\cdot0606, \\
\text{Equal temperament semitone} &= 1\cdot0595.
\end{aligned}$$

* Actually he was unacquainted with the concept of "frequency" and spoke in terms of wave-lengths. I have expressed his ideas in more modern language.

Thus the Pythagorean octave was made up of five equal tones and of two equal hemitones, which were rather less than half-tones.

The scale was perfect and complete—so far as it went. In addition to the concord of the octave, it contained no fewer than four fifths and five fourths, a greater wealth of concords than can be attained from any other selection of eight notes.

The scale could of course be extended indefinitely in either direction by a process of trespassing into neighbouring octaves. On the other hand, the severity of Greek taste resulted in melodies being restricted to a compass of an octave—and frequently even of a fourth—so as to employ only the best and most agreeable registers of the human voice, so that a trespass into an upper octave involved a corresponding curtailment of the lower octave. The normal eight-stringed lyre might begin at any note of the scale, but it would end at the same note in the octave above. As there were seven choices possible for the lowest note— c, d, e, f, g, a and b—this could be done in seven ways, which were referred to as "modes". These were as follows:

Range	Ancient Greek name	Scale (beginning with c)	Glarean's ecclesiastical name
c–c′	Lydian	c, d, e, f, g, a, b, c′	Ionian
g–g′	Ionian*	c, d, e, f, g, a, b♭, c′	Myxolydian
d–d′	Phrygian	c, d, e♭, f, g, a, b♭, c′	Dorian
a–a′	Aeolian†	c, d, e♭, f, g, a♭, b♭, c′	Aeolian
e–e′	Dorian	c, d♭, e♭, f, g, a♭, b♭, c′	Phrygian
b–b′	Myxolydian	c, d♭, e♭, f, g♭, a♭, b♭, c′	Locrian
f–f′	Syntolydian	c, d, e, f♯, g, a, b, c′	Lydian

* Or Hypophrygian when in another pitch.
† Or Hyperdorian when in another pitch.

Mediaeval music, and ecclesiastical music in particular, took over the Greek modes. Originally only four were recognised as "authentic", namely, d–d', e–e', f–f' and g–g', these having been approved by Ambrose, Bishop of Milan, in the fourth century. Then Pope Gregory the Great added four more, which were known as Plagal. Finally, in the sixteenth century Glarean (*Dodecachordon*, Basle, 1547) distinguished twelve modes, and assigned Greek names to them, although many of his identifications with the ancient modes were incorrect. Some of these twelve modes had never been used at all, having been found unsatisfactory from the outset, while others fell into disuse in the course of time. Then, sometime in the seventeenth century, musicians began to find only two modes entirely satisfactory; they became known as the major C D E F G A B C and the minor A B C D E F G A.

A collection of notes played in succession does not of itself constitute a melody which can awaken our musical imagination; to satisfy modern musical feeling, there must be a further element, which we describe as tonality. Our musical thought does not wish to wander indifferently all over the scale; it remains associated always with one particular note, the tonic or key note, which we somehow think of as giving a fixed and central point. Just as the traveller thinks of each point of his journey in terms of its distance from his home, so we moderns think of each note of a melody in terms of its interval from the key note. The skilful composer contrives to make us conscious of the key note from the very beginning of his music, and keeps our minds conscious of its position through all the notes that are played. In general, for instance, we expect the music—or at least the bass of it—to end on the key note, just as the traveller expects his journey to end at his home; we refuse to accept any other ending place as final. Even ancient Greek music had a sort of key note—the tone of the middle string of the lyre; Aristotle tells us that "All good melodies often employ the tone of the middle string, and good composers often come upon it, and if they leave it, recur to it again; but this is not the case with any other tone."

At first the composer could give variety to his music by writing in many different modes, but as the number of available modes decreased, he found it necessary to impart variety in other ways. The modern musician not only writes his music in a great variety of keys, but also, to maintain the interest at a high level, finds it necessary to change frequently from one key to another in the same piece. He may begin in the key of C, using as his scale the sequence of notes

C D E F G A B C,

and may very soon change to the key of G, in which his scale consists of the notes

C D E F♯ G A B C.

We can now see the first great objection to the Pythagorean scale from the standpoint of modern music—it is impossible to modulate from one key to another because the scale contains no F♯, and indeed no semitones at all. And it was impossible to create them by halving the whole tone intervals, because two hemitones did not make a tone.

A second, and hardly less weighty objection, remains. The only numbers which enter into the frequency ratios of the Pythagorean scale are 2 and 3, with their powers $2^2 = 4$, $2^3 = 8$, $3^2 = 9$, etc.; the numbers 5, 7, 11 do not occur at all. But the frequency ratios of a note and its various harmonics are represented by the complete sequence of numbers 2, 3, 4, 5, 6, 7, ..., so that most of these harmonics are not represented by notes on the Pythagorean scale at all. And Helmholtz's theory of dissonance makes it clear that the pleasurable consonances are the harmonics, and not the notes of the Pythagorean scale.

To take the simplest instance, the fifth harmonic of C has five times the frequency of the fundamental. But the nearest note on the Pythagorean scale, e'', has a frequency $\frac{81}{16}$ or 5·06 times that of the fundamental C, and so is about a fifth of a semitone out of tune with the fifth harmonic of C. If we sound C, this latter insists on sounding anyhow, for we have seen (p. 83) that the natural harmonics alone are forced by resonance, and makes considerable discord with the Pythagorean e''. Thus the note that our ear wants to hear sounding with C is not the Pythagorean e but the harmonic e.

It will be clear from what has already been said that these complications are absolutely fundamental; they arise out of the laws of arithmetic, which the musician is completely powerless to alter. If we visited another planet, we should find the same laws there as on earth. Here, 3^{12} is very nearly equal to 2^{19}, which means that 12 fifths are very nearly equal to 7 octaves, and the same would be true there, so that if the inhabitants were at about the same musical level as we are, we might expect to find them employing the same diatonic scale as ourselves. But, there as here, complications would arise from the two numbers not being exactly equal; they like ourselves would have a "comma of Pythagoras", and like our own musicians might have devoted a vast amount of thought to minimising its baneful effects.

The Mean-Tone Scale

A solution which prevailed for many centuries led to a scale known as the "mean-tone" scale. Four steps from C on our clock-face, C–G–D–A–E, bring us to E, a third above C. We have already seen that the frequency ratio between E and C is 5·06 on the Pythagorean scale, whereas the pleasurable ratio is 5·00. The bearings of the mean-tone system were laid by diminishing each of the four steps C–G, G–D, D–A and A–E equally by such an amount as gave the exact frequency ratio 5·00 to the interval C–E. Each of these steps was accordingly represented by a frequency

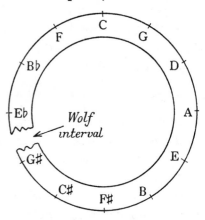

Fig. 56. The clock-face on the mean-tone scale.

ratio of $\sqrt[4]{5}$ or 1·49527, in place of the ratio 1·5 which characterises the exact fifth. The two quantities differ by about 3·15 parts in 1000. Each "hour" of the clock-face was accordingly in error by one 3·15 parts in a thousand, and the accumulation of twelve such errors amounted to 38 parts in a thousand, which is well over three-fifths of a semitone. When the scale was laid out as in fig. 56, the interval G♯–E♭ proved to be one of 7·395 equal temperament semitones, which is three-eights of a semitone more than the exact fifth (7·020 equal temperament semitones) —it was accordingly known as the *quinte-de-loup* or "wolf-fifth", wolves being howling animals.

Howling effects such as this could only be kept out of the music by carefully choosing the key in which a composition was written. By making slight departures from the mean-tone system, it was found possible to tune the notes so that music played in one key should sound harmonious, while that in a few other and nearly related keys did not sound too bad. For the rest, musicians simply had to avoid writing or playing in the more remote keys; they were virtually limited to three sharps or two flats, unless indeed instruments were specially arranged for them. Organs were sometimes built in which two black keys were interpolated between D and E, one sounding D♯ and the other E♭; and other notes were often treated in the same way. For instance, the organ which was built for the Foundling Hospital by Thomas Parker in the year 1768, "upon the new principle invented by the late Doctor Smith (Master of Trinity College, Cambridge)", contained devices for replacing the C♯, G♯, E♭, B♭ pipes by others sounding D♭, A♭, D♯ and A♯, the scheme accordingly being that shewn in fig. 57.

The general principles of the mean-tone system were foreshadowed by Schlick (*Spiegel der Orgelmacher und Organisten*, 1511), who suggested tuning the fifths F C, C G, G D, D A "as flat as the ear could endure", so that the third F A should "sound decent". In its more precise form the system seems to have been the invention of a blind Spanish musician Francis Salinas who lived the greater part of his life in Italy, and described the exact mean-tone system in his *De Musica Libri Septem* (1577). Gradually, but only very gradually, it was superseded by the system of "equal temperament", which had been proposed at an even earlier date, 1482, by another Spaniard, Bartolo Rames.

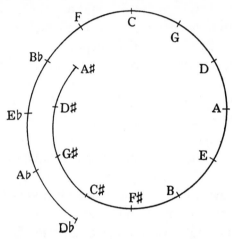

Fig. 57. The clock-face on the mean-tone scale, with four extra notes added, to make it possible to play in several keys.

The Equal-Temperament Scale

In this system the "comma of Pythagoras" is distributed equally over the twelve intervals which make up the circle on the clock-face. As the comma is about a quarter of a semitone, this involves flattening each interval of a fifth by about a forty-eighth of a semitone. Or to be more precise, since the twelve steps round the clock-face are to represent an interval of exactly seven octaves, and so a frequency ratio of 128 : 1, each step must represent a frequency ratio of $\sqrt[12]{128}$ or 1·4983. All semitones are now equal, and, as already explained on p. 25, each represents precisely the same frequency ratio, 1·05946. Although these frequency ratios had been correctly calculated by the French mathematician Mersenne, and published in his *Harmonie Universelle* as far back as 1636, the system does not seem to have been employed in practice until late in the seventeenth century, when it began to be used in North Germany. The first occasion on which we hear of its use is in the famous organ which Arp Schnitger built for S. Jacobi at Hamburg in 1688–92; this is said to have been tuned by the builder to something which at least approached to equal temperament. J. S. Bach subsequently advocated the system; not only were his own clavichord and harpsichord tuned to it, but he wrote the well-known "forty-eight" (*Wohltemperiertes Klavier*) to prove that it enables compositions in all keys to be played without disagreeable discords. Yet even he was unable to convert contemporary organ-builders to the new system, and there seems to be no doubt that the organs of his day were usually tuned to the meantone system. This doubtless explains why Bach seldom wanders into the more remote keys in his organ works, in striking contrast to his compositions for the clavichord.

After the death of J. S. Bach, his son Philipp Emanuel Bach started an active campaign in favour of equal-temperament tuning, but its adoption was slow, and especially so in England; it was not until about the middle of the nineteenth century that English pianos began to be tuned to equal temperament, and not a single one of the English organs shewn in the Great Exhibition of 1851 was so tuned.

The equal-temperament system is now in universal use for keyed instruments, and has the great advantage that music can be played equally well in all keys. On the other hand its defects are many. The most obvious is that of all the seventy-eight intervals that lie within the range of a single octave, not a single one is in perfect tune; every one could be improved if there were not the others to think about. The pianist and organist accept this accumulation of lesser evils in order to escape the major evils of badly discordant intervals. But the violinist and singer are under no such necessity; as each interval comes along, they can make it what they like, and so naturally tend to make it that which gives most pleasure to the ear. Observations shew that the intervals which such performers produce when they are left to themselves differ greatly from those they produce when accompanied by an instrument tuned to equal temperament.

Just Intonation

An attempt to standardise the former intervals has led to the introduction of yet a further system, known as the system of "just intonation". This is limited to one single key, and aims at making the intervals as accordant as possible with both one another and with the harmonics of the key note and of the closely related tones.

The ratios chosen are shewn in the following table:

Note		C	D	E	F	G	A	B	C
Ratio of	to C	I	$\frac{9}{8}$	$\frac{5}{4}$	$\frac{4}{3}$	$\frac{3}{2}$	$\frac{5}{3}$	$\frac{15}{8}$	2
frequency	to preceding note		$\frac{9}{8}$	$\frac{10}{9}$	$\frac{16}{15}$	$\frac{9}{8}$	$\frac{10}{9}$	$\frac{9}{8}$	$\frac{16}{15}$	

It will be seen that most of the frequency ratios can be expressed in terms of comparatively small numbers, indicating consonant harmonies.

On the other hand the whole tones are not all equal,

some, known as major tones, having a frequency ratio of $\frac{9}{8}$ (1·125), while others, known as minor tones, have a frequency ratio of only $\frac{10}{9}$ (1·111). The two semitones have the same frequency ratio of $\frac{16}{15}$, but this is more than half the frequency ratio of any full tones, since $\left(\frac{16}{15}\right)^2 = 1·138$. The second column of the following table shews the intervals of just intonation for the scale of C;

Frequency ratios in scale

Note	Just intonation						Mean tone	Equal temperament
	Scale of C	Harmonics of						
		C	G	D	F	A		
C	1·000	1·000	1·031	0·985	1·000	1·042	1·000	1·000
C♯	—	—	—	—	—	—	1·045	⎫ 1·059
D♭	—	—	—	—	—	—	1·070	⎬
C×	—	—	—	—	—	—	1·092	⎭
D	1·125	1·125	1·125	1·125	—	1·146	1·118	1·122
E♭♭	—	—	—	—	—	—	1·145	⎫
D♯	—	—	—	—	—	—	1·168	⎬ 1·189
E♭	—	—	—	—	1·167	—	1·196	⎭
E	1·250	1·250	—	1·266	—	1·250	1·250	1·260
F♭	—	—	—	—	—	—	1·280	
E♯	—	—	—	—	—	—	1·306	⎫ 1·335
F	1·333	1·375	1·312	—	1·333	—	1·337	⎭
F♯	—	—	—	1·406	—	—	1·398	⎫
G♭	—	—	—	—	—	—	1·431	⎬ 1·414
F×	—	—	—	—	—	—	1·460	⎭
G	1·500	1·500	1·500	—	1·500	1·458	1·495	1·498
A♭♭	—	—	—	—	—	—	1·531	⎭
G♯	—	—	—	—	—	—	1·563	⎫
A♭	—	—	—	—	—	—	1·600	⎬ 1·587
G×	—	—	—	—	—	—	1·633	⎭
A	1·667	—	1·687	1·687	1·667	1·667	1·672	1·682
B♭♭	—	—	—	—	—	—	1·712	⎭
A♯	—	—	—	—	—	—	1·747	⎫
B♭	—	1·750	—	—	1·833	—	1·789	⎬ 1·782
B	1·875	—	1·875	—	—	1·875	1·869	1·888
C♭	—	—	—	—	—	—	1·914	
B♯	—	—	—	—	—	—	1·953	⎫
c′	2·000	2·000	2·062	1·969	2·000	2·083	2·000	⎬ 2·000

while adjacent columns give the harmonics of C, G, D, F and A on this scale, and also the intervals on the meantone and equal-temperament scales.

The table shews that, on the scale of just intonation, C, G and A are true harmonics of F; E, G, D of C; and B and D of G.

The frequencies given in the second column are those which would actually be played by a violinist playing in the key of C. If, however, his music modulates to the key of G, his A will no longer have $\frac{5}{3}$ times the frequency of C, but $\frac{9}{8}$ times the frequency of G, and so $\frac{27}{16}$ times the frequency of C; the frequency of his A changes from 1·667 to 1·687 times that of C. Thus the pitches of his notes are not fixed, but vary with the key in which he happens to be playing at the moment.

The classical observations on this subject are due to Delezenne and Helmholtz. The latter wrote:

That performers of the highest rank do really play in just intonation has been directly proved by the very interesting and exact results of Delezenne. This observer determined the individual notes of the major scale as it was played by distinguished violinists and violoncellists, by means of an accurately gauged string, and found that these players produced correctly perfect Thirds and Sixths. I was fortunate enough to have an opportunity of making similar observations by means of my harmonium on Herr Joachim. He tuned his violin exactly with the g d a e of my instrument. I then requested him to play the scale, and immediately he had played the Third or Sixth, I gave the corresponding note on the harmonium. By means of beats it was easy to determine that this distinguished musician used b_1 and not b as the major Third to g, and e_1 not e as the Sixth.

Key Characteristics

On an instrument tuned to equal temperament, the semi-tones are all equal, so that the scales which represent the different keys differ only in pitch. They are completely similar in all other respects, the frequency ratios being the same in all. We can verify this by making a gramophone record of a chromatic octave C, C♯, D, D♯, E, ... C, played on a piano or other instrument tuned to equal temperament, and running it through the gramophone at 1·05946 times the speed at which it was taken. Then C becomes C♯ exactly, the C♯ becomes D exactly, and so on, so that what we hear is exactly the chromatic octave of C♯.

On an instrument which is not tuned to equal tempera-ment, the semitones are not all equal, so that a musician whose ear was infinitely sensitive would say: "This is not the chromatic octave of C♯ that I hear; it is the octave of C played a semitone too high." It follows that in every system of tuning other than equal temperament, each scale has its own special characteristic quality; we do not pass from one scale to another by a mere uniform change of pitch.

Some regard it as a defect of equal-temperament tuning that it obliterates the different characteristic qualities of the various keys. How serious this defect is will, of course, depend on how far these differences, if indeed they can be perceived at all, contribute to the interest or enjoyment of our music.

In the original Greek modes, the octave was divided into its seven intervals by steps which varied greatly from one

mode to another, with the result that the characteristic
qualities of the various modes were unmistakable and could
be recognised at once. Plato tells us, for instance, that the
Lydian mode (our modern major mode!) was specially associated with sorrow; it and the closely associated Ionian
mode, which only differed from it in b♭ replacing b, were
also the modes of softness, relaxation, self-indulgence, and
even drunkenness. The Dorian and Phrygian modes on the
other hand were—so he tells us—associated with courage,
the military spirit, temperance and endurance. Because
of this association, Plato would have permitted only the
Dorian and Phrygian modes to be employed in his ideal
republic, the Lydian and Ionian modes being prohibited.

The only modes which are in general use in modern
music are the Greek Lydian (major mode) and Aeolian
(minor mode), and their characteristic qualities are still
easily recognisable. There was a time when the church
frowned on the major mode as being too sensual for ecclesiastical music, but to-day we associate the major mode
primarily with strength, virility, gaiety, and even frivolity,
while the minor mode suggests sadness, seriousness and
profundity; indeed, because of these associations, such ex-
pressions as "in a minor key" are in common use as part of
the English language.

The differences between the various major keys are
far more subtle than those which differentiated the various
Greek modes, or those which produce the differences
between the major and the minor keys in modern music.
Instead of depending on the difference between whole
tones and semitones, they depend at most on the difference
between major whole tones and minor whole tones (p. 177).

When we compare two scales in major keys with one another, we find that, unless the tuning is that of equal temperament, the octave is still divided into its seven intervals by slightly different steps, and the question is whether this slight difference is perceptible to the trained musical ear, and if so, whether it has an appreciable influence on the emotional qualities of the music.

Many musicians, including Berlioz, Schumann and Beethoven, seem to have believed that both questions must be answered in the affirmative. We find Beethoven writing of B minor as a "schwarze tonart", describing Klopstock as "always *maestoso*—Db major", changing the key of a song in an effort to make it sound *amoroso* in place of *barbaresco*, and so forth.

The scientific Helmholtz appears to have held similar views; he wrote:

There is a decidedly different character in different keys on pianofortes and bowed instruments. C major and the adjacent Db major have different aspects. The difference is not caused by a difference of absolute pitch, as can easily be verified by comparing two instruments which are tuned to different pitches. If Db on one instrument has the same pitch as C on the other, the C major still retains its brighter and stronger character on both, and the Db its soft, veiled harmonious quality.

To-day many musicians claim to hear the different characteristics very clearly, and associate them with the emotional quality of the music. They will tell us that music played in the "open" key of C major—with neither flats nor sharps in the key signature—sounds strong and virile; played in the key of G, with one sharp, it sounds brighter

and lighter; in D, with two sharps, even more so; and
so on. Every additional sharp in the key signature is supposed to add to the brightness and sparkle of the music,
while every flat contributes softness, pensiveness, and even
melancholy. Some writers go into greater detail. Here,
for example, is part of an arrangement suggested by Ernst
Pauer and quoted in the English translation of Helmholtz's *Tonempfindungen*:

C major—Pure, certain, decisive; expressive of innocence, powerful resolve,
　　　　manly earnestness and deep religious feeling.
D♭ major—Fullness of tone, sonority and euphony.
E major—Joy, magnificence, splendour; brightest and most powerful key.
E minor—Grief, mournfulness, restlessness.
F major—Peace, joy, light, passing regret, religious sentiment.
F minor—Harrowing, melancholy.
F♯ major—Brilliant, very clear.
G♭ major—Softness, richness.

It is clear that even if these qualities had ever been
associated with the various keys, they must all be lost in
equal temperament, in which, to take the most obvious
instance, the key of F♯ major (six sharps) is absolutely
identical with that of G♭ major (six flats). Yet musicians
who claim to find the association in music played by the
orchestra claim also to hear it on the pianoforte, and have
expended much ingenuity in maintaining that the keys
retain their alleged characters and distinctive quality
even on the pianoforte. Helmholtz, for instance, argued that
the operation of striking a short black key must necessarily
differ mechanically from that of striking the white keys
with their longer leverage, and suggested that this may
cause the required difference. It cannot be denied that it
might make some difference, but it would be a most amaz-

ing coincidence if it made precisely the difference needed, so that the different lengths of the black and white keys on the pianoforte gave just the same characteristics to music played on the pianoforte as the deviations from equal temperament give to the same music when played by an orchestra.

Such a coincidence is, indeed, so utterly improbable that it seems safe to rule it out, and to assert that, in the case of pianoforte music at least, the special qualities of individual keys exist only in the imagination of the hearer, and possibly sometimes in that of the composer also, who may have chosen the key of a particular composition so as to fit in with his preconceived ideas of its emotional characteristics. In confirmation of this it may be remarked that pianoforte pieces retain their emotional qualities when played on a pianola, on which the mechanical difference between white and black keys disappears entirely.

The case of orchestral music cannot be dismissed so easily, but an obvious judgment may reasonably be based on the circumstance that those who claim to hear differences in the orchestra claim to hear precisely the same differences on the piano, although the equal temperament tuning makes it impossible that they should occur; and we may feel confirmed in our judgment by the circumstance that these differences are not always the same with all hearers. We have already noticed how Plato associated our modern key of C major with sorrow, weakness and self-indulgence, while Helmholtz associates it with brightness and strength, and Pauer with purity, innocence, manliness, and other virtues. And Helmholtz, in the passage just quoted, describes D♭ major as soft, veiled and harmonious, while Beethoven associated it with *maestoso* qualities, and Pauer's list tells us that it has fullness of tone, sonority and euphony.

All this suggests that the whole matter is one of subjective imagination, possibly based in the first instance on association of ideas. An obvious chain of associated ideas starts from sharps in the key-signature, and runs through sharpening of pitch to high notes and bright, joyous music; another runs from flats through flattening of pitch to deep-pitched notes, with their depression and seriousness. Obviously this does not explain everything; there may also be association with well-known pieces of music.

The power of subjective imagination seems to be very strong. Some hearers even claim to find emotional qualities in individual notes—here is a list from Curwen's *Standard Course of Lessons and Exercises in the Tonic Sol-fa Method* (1872):

> Do (key-note)—Strong, firm.
> Re—rousing, hopeful.
> Mi—steady, calm.
> Fa—desolate, awe-inspiring.
> So—grand, bright.
> La—sad, weeping.
> Ti—piercing, sensitive.

We cannot but be reminded of the Beethoven enthusiast who claimed that a single chord, nay even a single semiquaver, of his favourite master contained more emotional quality than all the music of Bach added together.

In whatever way we answer these various questions it remains true that the introduction of equal temperament tuning has resulted in much of the music of the earlier masters not being heard tonally as it was intended to be heard—as for instance the vocal and organ works of Bach and Handel, and the clavichord and harpsichord works of Handel, all of which were written for the mean-tone system.

Our discussion will have made it clear that there is no perfect system of intonation, and that no scale can be devised which is suited for all instruments. In an orchestra we may hear the brass playing in harmonics, the strings in just intonation, or perhaps a compromise between this and equal temperament, and the organ, harp and piano in equal temperament. Yet we seldom feel that anything is wrong, except perhaps in a pianoforte concerto, where the conflict arises in its acutest form. It was not always so. Dr Robert Smith, writing in 1759, described equal temperament as "that inharmonious system of 12 hemitones, which produces a harmony extremely coarse and disagreeable", and even in 1852 Helmholtz wrote:

When I go from my justly-intoned harmonium to a grand pianoforte, every note of the latter sounds false and disturbing....On the organ, it is considered inevitable that, when the mixture stops are played in full chords, a hellish row must ensue, and organists have submitted to their fate. Now this is mainly due to equal temperament, because every chord furnishes at once both equally-tempered and justly-intoned fifths and thirds, and the result is a restless blurred confusion of sounds.

While we cannot deny the general truth of this, we hardly feel so critical to-day. Perhaps our ears are more tolerant than those of our ancestors. Just as we have learned to tolerate and even enjoy harmonies which they found unbearable, so we may have learned to enjoy imperfectly tuned intervals which they heard only as a "hellish row".

The Music of the Future

Earlier in the present chapter we let our fancy roam to the extent of imagining that we were visiting another planet, on which musical development had reached about the same level as on our own. As the laws of arithmetic would be the same on this planet as on earth, we conjectured that the inhabitants might quite possibly have arrived at the same musical scale as our own, the octave being divided into twelve equal, or approximately equal, divisions.

If we are prepared to take a further flight of fancy, let us imagine that we visit a planet on which music has developed to a far higher level than our own, or, if we take an optimistic view of the future of our race, let us imagine that we revisit our own planet some thousands of years hence. What kind of music shall we find and, in particular, what scale will be in use?

The simplest, although not necessarily the correct, conjecture is that the music of the future will be like that of the present, but intensified—as it is now, only more so. To see what is implied in this, we must read our histories of music and imagine that those tendencies which have moulded music into its present form persist, and mould it still further in the same direction.

One tendency is typified in the history of consecutive fifths. Harmonies which have seemed venturesome and perhaps ugly to one generation seem natural and beautiful to the next, but are destined through continued repetition to seem obvious and tedious to generations yet to come. The sated ear for ever demands new harmonies which it will fast learn to tolerate, and then dismiss as threadbare and uninteresting.

Thus we find a long succession of musicians, Palestrina, J. S. Bach, Beethoven, Liszt, Wagner, Debussy—each of whom broke new ground, and most of whom were regarded as revolutionaries in their day—and innumerable other modern composers, introducing chords which after being thought perilously discordant at first, have now passed into the common language of music, and are heard with pleasure by our modern ears. Not only so, but the indisputable dissonances of equal temperament no longer distress us in the way that they seem, from the quotations given above, to have distressed our more fastidious predecessors.

One way of picturing the future is to imagine our posterity becoming more and more tolerant of dissonance as time goes on. If they ever attain to a stage in which all possible combinations of notes in the present scale are heard as tolerable but boring concords, further progress will only be possible by music enlarging its territory—by adding more notes to the scale. There is already a tendency to experiment with split semitones—quarter-tones—although up to the present it can hardly be said to have met with overwhelming success.

This brings us to a second tendency in musical history—a tendency continually to enlarge the scale. This has been in turn pentatonic (5-note), heptatonic (7-note) and chromatic (12-note). Has it reached its final resting place in the 12-interval division of the octave, or will the subdivision still continue?

We have already seen that the question is one for the arithmetician. Without forgetting the proverbial dangers of prophecy, we may be fairly sure that the laws of arithmetic will not alter, and that the natural harmonics will not change their position—a million years hence, as now, their frequencies will be 2, 3, 4, . . . times that of the fundamental. And, unless the physiological quality of our ears changes appreciably, we may assume that we shall always obtain our basic pleasure from chords whose frequency ratios can be expressed by the smallest of numbers. Because of this, it seems likely that the present fifth, with the simplest frequency ratio of all, 3 : 2, and the major third, with the next simplest frequency ratio 5 : 4, will figure largely in the music of the future. Before we attempt a conjecture about the musical scale of the future, it is worth seeing how far the subdivision of the octave would have to be extended, to provide a scale richer and purer in this respect than our present scale.

More Complex Scales

We have already seen that the complexities of the present scale centre in the fact that 12 fifths are not exactly equal to 7 octaves. Let us first examine whether we can replace the numbers 12 and 7 in this statement by others which will reduce the degree of inexactness. Using the method of continued fractions, we find that the following are increasingly good approximations to the ratio of the intervals of a fifth and an octave:

$$
\begin{array}{llllll}
12 \text{ fifths} = & 7 + 0.019 \text{ octaves} = & 7 \text{ octaves} + & \tfrac{1}{4} \text{ semitone} \\
41 \quad \text{,,} \quad = & 24 - 0.016 \quad \text{,,} \quad = 24 & \text{,,} & - \tfrac{1}{5} & \text{,,} \\
53 \quad \text{,,} \quad = & 31 + 0.003 \quad \text{,,} \quad = 31 & \text{,,} & + \tfrac{1}{28} & \text{,,} \\
306 \quad \text{,,} \quad = & 179 - 0.0014 \quad \text{,,} \quad = 179 & \text{,,} & - \tfrac{1}{60} & \text{,,}
\end{array}
$$

Each of these approximations is the best that can be obtained without extending the scale beyond the number of notes it contemplates, so that if the only problem was that of reducing the comma of Pythagoras to a minimum, the logical stopping-places would be at 12, 41, 53 and 306 notes to the octave. This is, however, far from being the whole problem: we want a scale which is rich in 5:4 consonances (major thirds) as well as in 3:2 consonances. Now in the various scales just mentioned, the number of notes which constitute the exact 5:4 consonance are found to be 3·86, 13·20, 17·06 and 98·51 respectively. The only scale which is even as good as the present 12-note scale in this respect is the 53-note scale. On this the present "fifths" are replaced by intervals of 31 notes, the tuning being almost perfect, while the present "major thirds" are replaced by intervals of 17 notes, these being flat by only a seventieth part of a present semitone.

So far as is known, a 53-note scale was first proposed in Europe in the seventeenth-century by Nicolas Mercator, Danish mathematician and astronomer, who, according to Yasser, found it mentioned in the writings of a Chinese theorist, King-Fang, of the second century B.C. In the middle of the last century, two harmoniums with 53 notes to the octave were built, one for Mr R. H. M. Bosanquet of London, and one by Mr J. P. White of Springfield, Mass., but neither seems to have been regarded as more than a curiosity.

We have already seen that the present 12-note scale has its roots embedded very deeply in the unalterable properties of numbers; we now find that music will have to go very far before finding a better scale. But a 53-note scale would give far purer harmonics than the present scale, and we can imagine future ages finding it worthy of adoption, in spite of all its added complexities—especially if mechanical devices replace human fingers in the performance of music. For, in the last resort, our limited scales have their origin in the limitation of our hands.

Yet, if ever music becomes independent of the human hand, may not the race then elect to use a continuous scale in which every interval can be made perfect—as with the unaccompanied violin of to-day?

CHAPTER VI

THE CONCERT ROOM

We have now considered the way in which musical sounds are produced by various instruments, and the way in which the quality of those sounds depends on the proportion of the various harmonics of which they are the blend. We have further considered the combining of single sounds into chords, and the choice of a musical scale which shall give as much pleasure as possible—or, perhaps, more accurately, as little pain as possible—to the sensitive ear. But our problem does not end with the production of musical tone; it ends only with the perception of this tone by the brain. After a musical tone has been produced in the orchestra, it has to go through two further stages— transmission from the instrument to the ear-drum of the listener, and transmission from his ear-drum to his brain. We consider these two stages in this and the following chapter.

The Transmission of Sound-Waves

Imagine that we are listening to an orchestra out in interstellar space. As we have seen that some material medium is necessary to transmit the sound from the instruments to our ear-drums, let us give free rein to our fancy, and imagine interstellar space filled with ordinary air.

Sound leaves each instrument in the form of waves which spread in all directions. If there is nothing to absorb their energy, they travel on for ever, but their intensity naturally diminishes as they spread out. One second after an instrument has been sounded, the waves that it has produced lie on a sphere 1100 feet in radius; after two seconds they lie on a sphere 2200 feet in radius, and so on.

The amount of energy which crosses these spheres is the same for all, but as the second sphere has four times as much surface as the first, only a quarter as much energy crosses it per square inch of its surface. If, then, one listener *A* listens at a distance of 1100 feet, and another *B* at a distance of 2200 feet, only a quarter as much energy will fall on the ear-drum of *B* as falls on the ear-drum of *A*. This is why the music sounds fainter to *B* than to *A*. If *A* gradually changes his distance from an instrument, the energy he receives will fall off in the same way, i.e. according to the law of the inverse square of the distance, which is the way in which gravitational force, or electrical attraction and repulsion, or the intensity of light, fall off—it is the natural law of diminution with distance.

Any single note of the music can be analysed into its harmonics in the way already explained, and there is no interchange of energy between the different harmonics as the sound travels through space. Thus the sound which falls on the ears of *B* is exactly identical in quality with that which falls on the ears of *A*; the only difference is in quantity.

Yet we must not assume that the music which *B* hears will be merely a fainter version of that which *A* hears. All sounds are treated in the same way until they fall on the ear-drum, their intensity falling off in the way just explained, whatever their pitch may be. But we shall see in the next chapter that the ear itself is not so impartial; it treats sounds of different pitches in different ways. For instance, when the sounds are faint, it is far more sensitive to treble notes than to bass, so that as the sounds of our interstellar orchestra fall off in strength, we are likely to hear treble notes far longer than bass notes. If the orchestra is a drum and fife band, the fifes will be heard to a far greater distance than the drums.

If we bring our orchestra down to earth, many new complications arise, of which three at least are important enough to be mentioned.

In the first place, the sound starts out as before in all directions, but some of it strikes the ground almost immediately. Waves of sound are subject to the same laws as other waves, so that when they strike the ground some of their energy enters the ground and is absorbed, while the remainder is reflected and starts off on a new path through the air, forming a sort of echo from the ground. Thus a listener may hear the sound of an instrument twice over, part of the sound having travelled directly through the air while the rest has first travelled to the ground, been reflected there, and then travelled up to the ear of the listener. This second part will arrive slightly later than the first, because it has travelled over a longer path, but the difference in time will usually be imperceptible, and the reflected sound will seem merely to reinforce the direct waves. Here again notes of different pitch are treated unequally, but the discrimination is now usually against the high notes; the sound of the drums is reflected better than that of the fifes.

In the second place, obstacles such as trees or buildings may make "sound-shadows", and so block out the sound from certain regions. Again notes of different pitch are treated unequally; the short sound-waves of a shrill note are completely stopped by a fair-sized obstacle, but those of a deep note merely swing round the obstacle, and re-unite again behind it. Indeed, this is a general property of waves of all kinds. We see it illustrated when sea-waves roll in and strike against the pillars of a seaside pier. Long waves seem to disregard the pillars altogether, merely dividing right and left and reuniting the moment the pillars are left behind. Short waves and small ripples, on the other hand, find the pillars a serious obstacle to progress; they are reflected back and spread out as new ripples in all directions. We may almost say that obstacles filter out short waves, while the long waves pass on undisturbed. We find the same thing occurring with light-waves. These are millions of times shorter than sound-waves, so that quite small obstacles, such as minute particles of dust, filter out the short waves which constitute blue light, and allow the long waves which constitute red light to pass. For this reason the sun looks red at sunset, or when seen through smoke or dust; the short waves which are filtered out make the blue of the day sky or the purple of the twilight. In every case the discrimination is against short waves, which in sound means tones of high pitch.

In the third place, a certain amount of sound will travel, not through the air, but through solid substances such as the floor or ground. Generally speaking, deep-pitched notes are transmitted more freely than those of higher pitch, so that once again the discrimination is against the latter.

Thus although high notes may be found to have the greater carrying power out in interstellar space, low notes will probably have the greater carrying power down on earth.

When we take our orchestra into a concert hall, the same complications recur in renewed strength. Walls, ceiling, furniture and audience are all at work absorbing part of the sound and reflecting the rest, while the columns, galleries and projections on the walls form shadows, and so obstruct the passage of the sound. Not only are sounds of different pitch treated unequally, but the different harmonic constituents of a single note are treated unequally, so that the proportions in which these are heard may vary as we pass from one point to another in a concert room. Finally, we have to reckon with sound being transmitted through solids as well as through the air.

All these complications provide problems for the architect, the builder and the conductor of the orchestra, who must try to secure that they act to the advantage, and not to the detriment, of the music. The whole subject is of great complexity, so that the most we can hope in the present book is to put the non-expert in a position to understand how the expert tackles his problems.

Reflection and Absorption of Sound

The most important problems, as well as the most interesting, are presented by the reflection and absorption of sound. The table overleaf shews the proportion of sound of various pitches which is absorbed by surfaces of various kinds.

The entries in the first line do not of course represent the results of experiments, but of common sense—when sound reaches an open window, 100 per cent. of it disappears;

none is reflected back. The entries in the other lines record the results of experiments made by Professor W. C. Sabine in America, and by various experimenters at the British Building Research Station and National Physical Laboratory.

Substance	Absorption of sound of pitch						
	CC	C	c'	c''	c'''	c^{IV}	c^V
Open window	1·00	1·00	1·00	1·00	1·00	1·00	1·00
Floor:							
Wood blocks, ¾ inch pine, laid in mastic	—	0·05	0·03	0·06	0·09	0·10	0·22
Pile carpet, ⅝ inch thick, on concrete	—	0·09	0·08	0·21	0·26	0·27	0·37
Rubber carpet, 3/16 inch thick, on concrete	—	0·04	0·04	0·08	0·12	0·03	0·10
Walls and Ceiling:							
18 inch brick wall unpainted	0·021	0·024	0·025	0·031	0·042	0·049	0·07
18 inch brick wall painted	0·011	0·012	0·014	0·017	0·020	0·023	0·025
Gypsum plaster or hollow tiles	0·012	0·013	0·015	0·020	0·028	0·040	0·050
Tiles (West Point)	0·012	0·013	0·018	0·029	0·040	0·048	0·053
Lime plaster on laths	0·048	0·020	0·024	0·034	0·030	0·028	0·043
Do, painted	0·036	0·012	0·013	0·018	0·045	0·028	0·055
Canvas, 6 inches from wall	—	0·10	0·12	0·25	0·33	0·15	0·35
Teak panelling, 3-ply, 1 inch from wall	—	0·09	0·17	0·17	0·15	0·15	0·15
Furniture:							
Hair cushion, under canvas and leatherette	0·25	0·42	0·47	0·72	0·47	0·27	0·16
Hair felt (12 per cent. solid)	0·09	0·10	0·20	0·52	0·71	0·66	0·44
Seated audience	0·35	0·72	0·89	0·95	0·99	1·00	1·00

The entries in the second line will serve to illustrate the uses of the table. We see that a wood-block floor absorbs 3 per cent. only of sound of pitch c' (middle C), but 22 per cent. of sound whose pitch is four octaves higher. All the sound which is not absorbed is reflected, so that a wood-block floor reflects 97 per cent. of c' sound, but only 78 per cent. of c^v sound. A simple calculation shews that after ten reflections the amounts of sound will be:

at c' 73 per cent. reflected; 27 per cent. absorbed
at c^v 8   ,,   92   ,,

while after twenty reflections the amounts are:

at c' 54   per cent. reflected; 46   per cent. absorbed
at c^v 0·7   ,,   99·3   ,,

To provide a simple illustration, let us imagine a room covered completely with wooden blocks—floor, walls, ceiling and furniture. Then the calculations just made shew that, in such a room, a note of pitch c' undergoes twenty reflections and more before being reduced to half its original intensity, whereas a note of pitch c^v is largely extinguished after about ten reflections, and almost completely so after twenty reflections. Thus a wood-block surface filters out the shrill components from sound. We can hardly expect to find a concert room completely furnished with $\frac{3}{4}$ inch pine blocks, but our table shews that other substances do the same thing to an even greater extent—carpets, felt and audiences in particular. We shall see later how effects such as these make a room bad for music, robbing the sound of its richness and quality.

As a second, and more realistic illustration, let us carry out similar calculations for a room in which the walls and floor are tiled, while the ceiling is of lath and plaster. Our table shews that these substances all behave in much the same way, average values for their absorption being:

at C absorption $= 1\cdot2$ per cent.
at c'   ,, $= 1\cdot8$,,
at c^v ,, $= 5\cdot3$,,

The amounts of sound after ten reflections are now:

at C 89   per cent. reflected; 11   per cent. absorbed
at c' 83   ,,   17   ,,
at c^v 58   ,,   42   ,,

and, after a hundred reflections,

at C 30   per cent. reflected; 70   per cent. absorbed
at c' 16   ,,   84   ,,
at c^v 0·4   ,,   99·6   ,,

Clearly this room is very different acoustically from the room we first considered. In the present room the major part of the sound of all pitches still persists after ten reflections; after a hundred reflections the shriller constituents have been filtered out, but tones up to c' still occur in considerable strength.

Reverberation

Even if the room has dimensions of only about ten feet, sound which has been reflected 100 times must have travelled about 1000 feet of path, and occupied nearly a second in the process. Thus any note below middle C will go on echoing round the room for at least a second before becoming silent. A bass or tenor voice will resound in all its richness, since its harmonics are only filtered out to a slight degree, but the same is not true for a soprano, hence the peculiarly male pleasure of singing in the bathroom.

Suppose next that the tiled and plastered room, instead

of being a mere bathroom, has dimensions of 100 feet in all directions. Again sound is extinguished only when it meets the walls, and all except the shriller constituents can be reflected about 100 times before extinction; by this time it will have travelled about 10,000 feet, and occupied about 9 seconds in so doing. We now have what is described as a very "reverberant" room. Both speech and music are quite impossible in it; speech because when any syllable is spoken the twenty or thirty preceding syllables are still echoing round the room, and music because every note is smeared out into a sustained note of many seconds' duration.

In designing a room for speech or music, it is clearly important to avoid too much reverberation. This can always be done by using fairly absorbent materials for the walls, etc., but there is then a danger that these will rob the music of its richness by absorbing the higher harmonics more than the lower (see p. 213, below).

Stretched strings and other similar old-fashioned artifices are probably quite worthless for the lessening of reverberation; at any rate no scientific justification for their use has yet been found. The coupling (p. 131) between stretched strings and the air of the room must always be far too loose for the strings to serve any purpose; if the strings were stroked with a violin-bow they would not fill the hall with sound, and this shews that they cannot absorb any appreciable amount of sound from the hall.

General Theory of Acoustics

The special detailed problems we have so far discussed are only instances of a wide general theory, due largely to Professor W. C. Sabine of Harvard, which we shall now

explain. Strictly speaking, the theory has reference only
to very reverberant rooms, but the results it gives are found
to be true, to a reasonably good approximation, for all
except very non-reverberant rooms.

We have seen that the music which a listener hears in
a reverberant room consists of different parts; one part
comes direct from the musical instrument, but other parts,
almost equal in intensity, reach him after one, two, three
or more reflections. The majority of the sound which enters
his ear has been reflected many times, and this being so, its
amount is much the same whether he is near the front or
the back of the hall, or anywhere inside it. Indeed, we
know that only the most inexperienced beginner chooses
the front seats in a concert under the impression that
he will hear the music best from there; the experienced
musician is more likely to select a seat far back, where the
instruments are all at about the same distance, and so are
heard in good balance. Any inadequacy in the volume of
sound is far more likely to result from obstructions which
cause sound-shadows (p. 193) than from mere distance.

Sabine's theory, at least in its simplest form, proceeds on
the supposition that the energy of sound is spread uniformly through the hall, so that every cubic foot of space
contains the same amount. This is of course accurately true
only in a very reverberant room. The energy does not stand
still, but travels in all directions at a speed of 1100 feet a
second, and this results in its extinction. A stream of sound
energy is continually falling on all the walls and pieces of
furniture in the room. If these were perfectly reflecting,
no energy could be absorbed, and the sound would stay
always at the same level of intensity. But, as no walls or

furniture can reflect perfectly, the sound is continually being absorbed, and finally dies away.

Let us nevertheless imagine that all the walls, floor, ceiling and furniture of a room are perfectly reflecting, so that there is no loss of sound by absorption here, but let us also suppose that the room has a window exactly one square foot in area, and that we suddenly open this. Sound at once streams out through the window, never to return, so that the total energy of sound in the hall begins to diminish and continues to diminish until none is left.

How long will this process take? For a preliminary computation we may picture a stream of sound energy pouring out of the window with a speed of 1100 feet a second, so that 1100 cubic feet of the room are cleared of energy every second. Clearly, then, if the room has a volume of *x* times 1100 cubic feet, the whole process will take *x* seconds.

This very rough-and-ready calculation needs amendment in two respects.

In the first place, we have reasoned as though the sound inside the room was all moving directly towards the windows; actually it is moving in all directions equally. Half of it is moving in directions which carry it farther away from the window, so that this half must not come into the calculation at all. And it can be shewn that the remaining half has an average speed *in the direction directly towards the window* of only one-half of 1100 feet a second, namely 550 feet a second. Thus the rate at which sound energy escapes through the window is not 1100 cubic feet a second, but only a quarter of this, namely 275 cubic feet a second.

In the second place, we have not allowed for the continual diminution in the energy left in the room. In every second the loss is the average energy in 275 cubic feet, but the average energy per cubic foot is a steadily diminishing quantity. We may still suppose the volume of the room to be x times 1100 cubic feet, which is $4x$ times 275 cubic feet, so that the energy now diminishes at the rate of one part in $4x$ of the whole per second. Now a man who each year spends one part in four of his remaining capital may become poor, but never penniless. In the same way, the energy in our room can never be reduced absolutely to zero. A simple calculation shews that it will be reduced to a millionth part of its value in 13·8 times $4x$ seconds, or $55\cdot2x$ seconds.

If we now write V for $1100x$, so that V is the number of cubic feet in the room, this time is very nearly equal to $\frac{1}{20} V$.

The time which sound takes to fall to a millionth of its original value is called the "period of reverberation" of the room. The choice of a factor of a millionth may seem somewhat arbitrary, but it is intended to represent the difference in intensity between a quite loud noise and one which we can just, but only just, hear. Roughly speaking, the "period of reverberation" is the time a loud sharp noise, such as a shout or hand-clap, takes to sink to inaudibility.

For our imaginary room, with perfectly reflecting walls and furniture and an open window one square foot in area, we have seen that this time is $\frac{1}{20} V$. With two open windows of this size, the sound escapes twice as fast, so that the "period of reverberation" is reduced to $\frac{1}{40} V$. If there are a whole lot of such open windows, n in number, the escape of sound is speeded up n-fold, and the "period of reverberation" is reduced to $\frac{1}{20n} V$. Or we can produce the same effect by opening a single window n square feet in area. We can cut down the period of reverberation of any room to as little as we like by opening sufficient windows—always assuming, of course, that the windows are there to be opened.

In an actual room, every square foot of wall absorbs energy, and so forms an outlet for the escape of energy just as an open window does. The only difference is that an open window absorbs all the energy that falls on it, whereas an area of wall absorbs only a fraction of the energy, namely that given in the table on p. 196. As the amount of this fraction depends on the pitch of the sound, we must imagine the sound analysed into its constituents of different pitches, and discuss these constituents separately. Let us suppose, for instance, that our sound is of pitch middle C, then we find that each square foot of unpainted brick wall absorbs a fraction 0·025, or a fortieth, of the sound which falls upon it, so that 40 square feet of brick wall have the same absorbing power as one square foot of open window.

Let us speak of the absorbing power of a square foot of open window as a "unit of absorption". Then we can pass round our room, and measure the number of units of absorption provided by the floor, ceiling, walls and furniture of the room in turn. If the total number of units proves to be n, the total absorption of sound in the room is equivalent to that of an open window n square feet in area, and the period of reverberation is $\frac{1}{20n} V$, where V is the volume of the room in cubic feet.

For sound of moderate pitch, lath and plaster, wood

floors, and glass windows all have about the same absorbing power as brick walls, namely one-fortieth, so that an empty room has a total absorbing power equal to about a fortieth of its total surface in square feet. If the dimensions of the room are f feet in all directions, the total surface is $6f^2$, and the total absorbing power is one-fortieth of this, or $\frac{3}{20}f^2$. As the volume is f^3 cubic feet, the formula $\frac{1}{20n}V$ shews that the period of reverberation is $\frac{1}{3}f$ seconds— one second for every yard in the linear dimensions of the room. The larger a room, the greater its reverberation.

Acoustical Analyses

As a first example of the use of our formula, let us study the acoustics of an exceedingly reverberant building—the Baptistry at Pisa. Except for its windows, the interior is almost entirely of marble, with an absorption coefficient of only about 0·01. The floor is circular with a diameter of 100 feet, and the roof is conical with an extreme height of 179 feet. Thus the interior has a total surface of about 50,000 square feet, and a volume of about 1,000,000 cubic feet. If the interior surface were wholly of marble, the floors, walls and ceiling would provide only about 500 units of absorption, so that the room would have a reverberation period of 100 seconds—sound would persist for more than a minute and a half. Under actual conditions, the observed reverberation period is 11 or 12 seconds. In this room, a man may sing a sequence of notes staccato, and hear them combined into a chord for many seconds afterwards.

A more detailed calculation for a small concert hall 50 feet long, 20 feet wide and 20 feet high, entirely empty of both furniture and audience, might stand as follows:

	Substance	Area (in sq. ft.)	Units of absorption				
			C	c′	c″	c‴	c^iv
Floor	¾ inch pine blocks in mastic	1000	50	30	60	90	100
Walls	3-ply teak	2800	252	476	476	420	420
Ceiling	Lath and plaster	1000	20	24	34	30	28
Total units of absorption (n)			322	530	570	540	548
Period of reverberation in seconds $\left(\dfrac{1000}{n}\right)$			3·1	1·89	1·75	1·85	1·82

If, instead of leaving the hall empty, we cover half the floor with an audience, we must replace the absorption of 500 square feet of pine blocks by that of an equal area of seated audience, and the calculation stands as follows:

	Substance	Area (in sq. ft.)	Units of absorption				
			C	c′	c″	c‴	c^iv
Floor	Seated audience (minus ¾ inch blocks)	500	335	430	445	450	450
Rest of hall	Brought forward	—	322	530	570	540	548
Total units of absorption			657	960	1015	990	998
Period of reverberation in seconds			1·52	1·04	0·99	1·01	1·00

The presence of the audience has reduced the period of reverberation to about half, as is often the case in actual practice (see p. 209, below), with the result, as we shall shortly see, that music sounds only half as loud as in the empty hall.

To take still a third example, let us consider the same room when the floor is covered with felt, and half the wall space is hung with tapestry or canvas. The calculation now stands as follows:

	Substance	Area (in sq. ft.)	Units of absorption				
			C	c′	c″	c‴	civ
Floor	Hair felt	500	50	100	260	355	330
,,	Audience	500	360	445	475	495	500
Walls	3-ply teak	1400	126	238	238	210	210
,,	Canvas	1400	140	168	350	462	210
Ceiling	Lath and plaster	1000	20	24	34	30	28
Total units of absorption			696	975	1357	1552	1278
Period of resonance in seconds			1·44	1·03	0·74	0·64	0·78

We notice that the introduction of felt and canvas has reduced the period of reverberation still further for sounds of high pitch, although leaving it approximately unaltered for sounds of low pitch.

Conditions for Good Hearing

Before we can discuss these results in detail, we must first consider another problem—the maintenance of a continuous sound in a room, such as might be produced from a steadily blown organ-pipe or a steadily bowed string. When such a sound first starts, the room may be supposed empty of sound. Gradually the sound fills the room, but as its intensity increases, the amount of sound absorbed by the walls also increases. Soon the sound intensity reaches a level beyond which it cannot pass, because at this level the walls are absorbing as much sound as the instrument is producing. If, for instance, the walls represent 1000 units of absorption, they will absorb 275,000 cubic feet of sound energy every second, and the level in question is reached when the sound energy is such that the instrument produces 275,000 cubic feet of it every second.

Let us next imagine that we halve the absorption of the room, as we might do by emptying the room of audience or by closing some windows. There are now only 500 units of absorption in place of 1000, so that when the steady state is reached our instrument is producing only 275 × 500, or 137,500, cubic feet of sound energy every second. Yet, as it is still the same instrument, it must produce just as much sound as before, whence we see that a cubic foot of space in the empty room must contain twice as much sound energy as it contained in the full room; now that the absorption is halved, the music sounds just twice as loud. If we had reduced the absorption to a quarter, the music would have sounded four times as loud, and so on. It is a general law that the loudness of a given instrument varies inversely as the number of units of absorption, and as the law is true for any and every instrument, it must of course be true for any orchestra or band or any collection whatever of noise-producing instruments. It is not only true as between different states of the same room, but also as between different rooms. If we move our instrument or orchestra to a new room, with only half as much absorption as the original room, it will sound twice as loud.

If we stay in the same room, the loudness varies inversely as the number of units of absorption, and as the period of reverberation also varies inversely as the number of units of absorption, it follows that the loudness of the music is proportional to the period of reverberation. If we pass from one room to another the loudness of the same instrument, or collection of instruments, is proportional to the period of reverberation divided by the volume of the room.

Thus we can always make music sound louder by lessening the number of units of absorption in a room, but in so doing we also increase the period of reverberation, with the result that the music sounds not only louder but also more blurred. Conversely we can make the music sound perfectly sharp and clear—in brief, perfectly un-blurred—by reducing the reverberation period of the room, but we can only do this by increasing the number of units of absorption, and so enfeebling the sound.

When the loudness of sound is unimportant—as for instance in broadcasting or recording for gramophones, where the sound can be amplified to any desired extent by electrical means—it is customary to use very absorbent walls, thereby attaining a short reverberation period and great clearness in the music. But music played under these conditions does not sound well to a listener in the room, and it becomes a question of balancing reverberation against clearness.

The Optimum Reverberation Period

Sabine, Watson, Lifshitz and others have found, from a series of careful experiments, that there is a certain "opti-mum" period of reverberation, at which practically all cultivated listeners agree that music sounds at its best. For reasons which are not clearly understood, this period is not the same for all rooms; it depends on the size of the room to a marked degree. To a lesser degree it depends also on the type of music, whether speech, song, light or serious instrumental music. The following table shews the periods of reverberation which are found to be best for the hearing of music in rooms of different sizes, the numbers given being the average of recent determinations by Lifshitz and Watson:

Volume of room (in cu. ft.)	Optimum period of reverberation (in seconds)	Number of units of absorption needed to give the optimum period
12,000	1·03	580
20,000	1·15	870
50,000	1·3	1,920
100,000	1·5	3,300
200,000	1·7	5,900
500,000	1·95	12,800
1,000,000	2·25	22,200

For comparison with the foregoing, the table which follows† gives the reverberation periods of a number of actual rooms and halls of which the acoustics are generally admitted to be really excellent:

Room	Seating capacity	Volume (in cu. ft.)	Period of reverberation		Optimum period for size, as given in foregoing table
			Empty	Full	
Lecture Theatre of the Royal Institution (London)	640	46,000	1·2	0·7	1·3*
Small Hall of the Conservatoire (Moscow)	550	90,000	3·46	1·30	1·5
House of Commons (London)	570	127,000	3·3	1·5	1·5*
Gewandhaus (Leipzig)	1500	400,000	3·6	2·3	1·9
Column Hall of the House of Unions (Moscow)	1600	443,000	3·55	1·75	1·9
Eastman Theatre (Rochester, N.Y.)	3340	790,000	4·0	2·08	2·1

* These rooms are excellent for speech, but untried for music. Watson finds that the optimum reverberation period for speech is only about 80 per cent. of that for music.

† This is compiled from material given in *The Acoustics of Buildings* by Davis and Kaye (Bell, 1932).

The Optimum Size of Orchestra

Closely connected with this question is that of the size of orchestra best suited for a room of given size. The inexperienced layman is likely to dismiss the problem as obvious. Doubling the volume of our hall, he will say, doubles the amount of space to be filled with music, and so calls for a doubling of the orchestra; hence he will want to make the orchestra proportional to the volume of the hall, or to the cube of its linear dimensions. The scientific musician, on the other hand, regarding units of absorption as mouths to be fed with sound, will see that the orchestra should be proportional to the total number of units of absorption. It ought, therefore, to be proportional to the surface, rather than to the volume, of the room—to the square rather than to the cube of its linear dimensions. In large halls, cathedrals and other very reverberant buildings, the greater part of the absorption will often come from the audience or congregation, so that it is almost permissible to say that the instrumental power ought to be proportional to the number of people in the building. The old organ-builder's rule of one pipe for every member of the congregation was no doubt a mere empirical rule-of-thumb, but it is easy to-day to find scientific justification for it.

If we wish to obtain the most pleasing results in any particular hall, we must remember that the ear tolerates, and even demands, more reverberation in a large hall than in a small one. The optimum period of reverberation for halls of various sizes is given in the second column of the table at the top of p. 209; it is easy to calculate from this how many units of absorption are needed to reduce the period of reverberation to its optimum value. The necessary number of units of absorption are given in the last column of the table.

Experiment shews that the average orchestral instrument can suitably be made to feed about 200 units of absorption; if it has more to feed, it will be unable to produce a satisfactory *fortissimo*, while if it has fewer to feed, the music will sound too overpowering when the instrument is played at its loudest. Thus to obtain the best results, an orchestra should contain one instrument for every 200 units of absorption, and to discover the best number of instruments for a hall of given size we must divide the numbers in the last column of the table on p. 209 by 200. The results are shewn in column two of the following table. Column three gives the size which is actually found best according to Heyl:

Volume (in cu. ft.)	Number in orchestra	
	Calculated	Heyl
12,000	3	—
20,000	4	—
50,000	10	10
100,000	17	20
200,000	30	30
500,000	64	60
800,000	90	90
1,000,000	111	—

In all these computations it has been assumed that the orchestra is one with a proper balance between the different classes of instruments, between the wind, string and percussion sections, between treble and bass, and so forth.

The Ideal Concert Room

We have so far been concerned only with the quantity of sound, but its quality is perhaps even more important. Two concert rooms may be exactly similar in size and shape, and yet music will sound rich, brilliant and full of life in the first, but dull, flat and dead in the second. In the first the performers play with zest and give of their best; in the second they feel a want of sympathy in their surroundings and a want of power in themselves. What is the cause of this difference?

In part it is a difference in the periods of reverberation of the two rooms. Since the loudness of the note produced by any instrument is proportional to the length of the period of reverberation, a long period naturally induces an exhilarating feeling of effortless power, not to mention a welcome slurring over of roughnesses and inequalities of force and tempo, while a short period produces the despair of ineffectual struggle: the music has only had time to shew its blemishes in all their nakedness, and is already dead. But further and important differences arise from the way in which the reverberation period changes as we pass up the scale of pitch.

The tables on pp. 205 and 206 contain acoustical analyses of two imaginary rooms of equal size. The former may be described as a "wood" room, the latter as a "felt and canvas" room. If we average over all pitches, the periods of reverberation of the two rooms are not widely different, and both are near to the optimum, 1·15 seconds, for a room of this size. If we do not average over all pitches, we notice a great difference in the way in which these periods vary with the pitch. In the "wood" room all notes above middle C have approximately the same reverberation period, so that high and low notes which are delivered with equal strength will fill the room with sounds of equal loudness; the balance between tones of different pitch is maintained. In the "felt and canvas" room, on the other hand, high tones have a specially low period of reverberation, and so must sound disproportionately thin and feeble, with the result that the music loses its richness and life. Something could perhaps be done towards remedying this by increasing the number and strength of the treble instruments, but the remedy is only partial. For it is not enough to take the notes of the various instruments as units; we must go deeper than this, and think of each note as a blend of harmonics. The "wood" room treats all harmonics above middle C equally, so that the timbre or characteristic quality of sound, which depends on the proportion in which the different harmonics are blended, does not suffer on reflection of the sound. But the "felt and canvas" room throttles the higher harmonics, and so alters the timbre of every note—usually for the worse. We have seen that the tone of a Stradivarius differs from that of a cheap modern violin mainly in possessing a superabundance of very high harmonics; the "felt and canvas" room will rob the Stradivarius of these constituents of its tone, and make it sound like a cheap violin. Again, eight harmonics at least are heard in oboe tone, the higher of these giving to the tone its thin, plaintive, acid-sweet quality. If the tone is robbed of these, it sounds fat and fluty. Indeed, every instrument in the orchestra becomes fat and fluty when robbed of its harmonics, with the result that all instruments sound more or less alike. The same is true of diapason, gamba and reed tone on the organ.

If, then, the music is to sound at its best, the interior of the concert room should be covered with substances which do not unduly deprive sound of its higher harmonics. Numbers of special absorbent materials and acoustic plasters and tiles are on the market which are designed to avoid this. As the table on p. 196 shews, ordinary wood panelling is not unsuitable for this purpose, its absorption being fairly uniform for notes of all pitches. If anything, it is unduly kind to tones of high frequency, and so actually enhances the brilliance of the music, a property which results in its frequent use.

It is especially important to use suitable materials for the ceiling and upper parts of the walls. Even in a fairly reverberant room, in which sound may be reflected tens, or even hundreds, of times before sinking below the threshold of audibility, the greater part of the sound we hear is still contributed by waves which have either come direct, or have been reflected only a very few times. To see this, let us imagine that the interior of our room is completely covered in material which absorbs one-third of the sound falling on it, and reflects the remaining two-thirds. In such a room sound is reflected no fewer than thirty-four times before it becomes inaudible, yet of the sound we hear, one-third comes from direct unreflected sound, and another two-ninths (making five-ninths in all) from sound which has been reflected only once. Sound which has been reflected more than three times contributes less than one-fifth of the whole. The most important contributions to the sound we hear are accordingly made by sound which has come direct and sound which has been only once or twice reflected. In an actual room the floor is often covered with highly absorbing material, and in part with a highly absorbing audience, so that the sound we hear consists largely of sound which has not yet struck the floor, and the reflecting qualities of ceiling and walls become of preponderating importance.

To make a room reverberant, its ceiling and walls ought to be especially massive, hard and smooth, since these, in a general way, are the properties which endow a surface with good sound-reflecting qualities.

Although the present book is not primarily concerned with this, similar considerations apply to the use of rooms for speaking. Our tables shew that a room has usually fewer units of absorption, and so is usually more reverberant for bass and tenor notes than for treble, so that less effort is needed to fill it with bass sound than with treble. In the matter of expenditure of energy, such a room favours the male speaker, but he must remember that he has to contend with a longer reverberation period than a woman speaker, and so must speak more slowly. It is often said that a room has a definite characteristic "pitch", and that a speaker can be heard with ease as soon as he adjusts his voice to the "pitch of the room". Everyone with much experience of public speaking must recognise that there is a certain amount of truth in this. The "pitch" probably represents some frequency for which the absorption is a minimum, and the reverberation period consequently a maximum. The explanation is frequently given that the pitch corresponds to one of the free vibrations of the air in the room, but this would seem to be untenable for innumerable reasons.

Quite apart from the materials used for its furniture and

decoration, the shape of a room must obviously have a great influence on its quality as a concert room. We have already seen that deep tones can travel round corners and pass round big obstacles far more freely than tones of high pitch. Again, then, the listener who is seated round a corner or behind an obstacle is in danger of hearing the music robbed of its higher harmonics. That room is best for music which has no obtruding corners or massive obstacles.

Various other practical points need careful consideration by the architect. For instance, a curved ceiling may focus sound reflected from it, so that it is heard in undue intensity at one point or along one line. To avoid this, care has to be taken that any such points or lines that there may be are either well above or well below regions which are likely to be occupied by the ears of listeners. A dome in the ceiling may be nearly as absorbent as an open window of equal area, since sound which once gets into it may only emerge again after a great number of reflections, which have so diminished its strength as to make it almost inaudible. And if it is still audible, it may have wasted so much time in the dome as to constitute merely a distracting echo, entirely detached from the main volume of sound. The same is true to a large extent of deep recesses and alcoves, and of narrow spaces in general.

The tendency of sound to find paths other than those through the air also needs careful consideration, but is not necessarily disadvantageous. In the Leipzig Gewandhaus the orchestra is placed on a raised platform which is deliberately connected by strong wooden beams to the panelling of the hall. In this way, the walls of the building are made to act as a huge sound-board, with excellent results.

CHAPTER VII

HEARING

We have now considered the generation of sound and its transmission through the air to the ear; we must finally consider its reception by the ear, and transmission to the brain.

When the air is being traversed by sound-waves, we have seen that the pressure at every point changes rhythmically, being now above and now below the average steady pressure of the atmosphere—just as, when ripples pass over the surface of a pond, the height of water in the pond changes rhythmically at every point, being now above and now below the average steady height when the water is at rest. The same is of course true of the small layer of air which lies in contact with the ear-drum, and it is changes of pressure in this layer which cause the sensation of hearing. The greater the changes of pressure, the more intense the sound, for we have seen that the energy of a sound-wave is proportional to the square of the range through which the pressure varies.

The pressure changes with which we are most familiar are those shewn on our barometers—half an inch of mercury, for instance. The pressure changes which enter into the propagation of sound are far smaller; indeed they are so much smaller that a new unit is needed for measuring them—the "bar". For exact scientific purposes, this is defined as a pressure of a dyne per square centimetre, but for our present purpose it is enough to know that a bar is very approximately a millionth part of the whole pressure of the atmosphere. When we change the height of our ears above the earth's surface by about a third of an inch, the pressure on our ear-drums changes by a bar; when we hear a fairly loud musical sound, the pressure on our ear-drums again changes about a bar.

The Threshold of Hearing

Suppose that we gradually walk away from a spot where a musical note is being continuously sounded. The amount of energy received by our ears gradually diminishes, and we might perhaps expect that the intensity of the sound heard by our brains would diminish in the same proportion. We shall, however, find that this is not so; the sound diminishes for a time, and then quite suddenly becomes inaudible. This shews that the loudness of the sound we hear is not proportional to the energy which falls on our ears; if the energy is below a certain amount we hear nothing at all. The smallest intensity of sound which we can hear is said to be at "the threshold of hearing".

We obtain direct evidence that such a threshold exists if we strike a tuning-fork and let its vibrations gradually die away. A point is soon reached at which we hear nothing. Yet the fork is still vibrating, and emitting sound, as can be proved by pressing its handle against any large hard surface, such as a table-top. This, acting as a sound-board, amplifies the sound so much that we can hear it again. Without this amplification the sound lay below the threshold of hearing; the amplification has raised it above the threshold.

In possessing a threshold of this kind, hearing is exactly

in line with all the other senses; with each our brains are conscious of nothing at all until the stimulus reaches a certain "threshold" degree of intensity. The threshold of seeing, for instance, is of special importance in astronomy; our eyes see stars down to a certain limit of faintness, roughly about 6·5 magnitudes, and beyond this see nothing at all. Just as a sound-board may raise the sound of a tuning-fork above the threshold of hearing, so a telescope raises the light of a faint star above the threshold of seeing.

We naturally enquire what is the smallest amount of energy that must fall on our ears in order to make an impression on our brains? In other words, how much energy do our ears receive at their threshold of hearing?

The answer depends enormously on the pitch of the sound we are trying to hear. Somewhere in the top octave of the pianoforte there is a pitch at which the sensitivity of the ear is a maximum, and here a very small amount of sound energy can make itself heard, but when we pass to tones of either higher or lower pitch, the ear is less sensitive, so that more energy is needed to produce the same impression of hearing. Beyond these tones we come to others of very high and very low pitch, which we cannot hear at all unless a large amount of energy falls on our ears, and finally, still beyond these, tones which no amount of energy can make us hear, because they lie beyond the limits of hearing.

The following table contains results which have recently been obtained by Fletcher and Munson.* The first two

* Many other investigators have worked at the problem, their results generally agreeing fairly closely, although not always exactly, with those stated in the table. The most recent investigations, by Andrade and Parker (1937), yield results which are in very close agreement with those of Fletcher and Munson.

columns give the pitch and frequency of the tone under discussion, the next column gives the pressure variation at which the tone first becomes audible, while the last column gives the amount of energy needed at this pitch in terms of that needed at f^{iv}, at which the energy required is least:

Tone	Frequency	Pressure variation at which note is first heard	Energy required in terms of minimum
CCCC (32-ft. pipe of organ; close to lower limit of hearing)	16	100 bars	1,500,000,000,000
AAA (bottom note of piano)	27	1 bar	150,000,000
CCC (lowest C on piano)	32	$\frac{2}{6}$,,	25,000,000
CC	64	$\frac{1}{40}$,,	100,000
C	128	$\frac{1}{200}$,,	3,800
c' (middle C)	256	$\frac{1}{1000}$,,	150
c''	512	$\frac{1}{2500}$,,	25
c'''	1,024	$\frac{1}{5000}$,,	6
c^{iv}	2,048	$\frac{1}{10000}$,,	1·5
f^{iv} (maximum sensitivity)	2,734	$\frac{1}{12800}$,,	1·0
c^{v} (top of piano)	4,096	$\frac{1}{10000}$,,	1·5
c^{vi}	8,192	$\frac{1}{2000}$,,	38
c^{vii}	16,384	$\frac{1}{100}$,,	15,000
Close to upper limit of hearing	20,000	500 bars	38,000,000,000,000

We see that the ear can respond to a very small variation of pressure when the tone is of suitable pitch. Throughout the top octave of the piano, less than a ten-thousand-millionth part of an atmosphere suffices; as already mentioned, this is produced by an air-displacement of less than a ten-thousand-millionth part of an inch, which again is only about a hundredth part of the diameter of a molecule.

We also notice the immense range of figures in the last column. Our ears are acutely sensitive to sound within the top two octaves of the piano, and quite deaf, at least by comparison, to tones which are far below or above this range; to make a pure tone of pitch CCCC audible needs a million million times more energy than is needed for one seven octaves higher.

The structure of an ordinary organ provides visual confirmation of this. The pipe of pitch CCCC is a huge 32-foot monster, with a foot opening which absorbs an enormous amount of wind, and yet it hardly sounds louder than a tiny metal pipe perhaps three inches long taken from the treble. A child can blow the latter pipe quite easily from its mouth, but the whole force of a man's lungs will not make the 32-foot pipe sound audibly.

Two Psychological Laws

So much for the smallest amount of sound that can affect our ears; let us next examine what is the smallest *difference* of sound that can affect them—in other words, by how much must a sound be increased before our ears inform us that it has become appreciably louder?

To answer this question, let us experiment with two electrically driven tuning-forks or loudspeakers, each provided with an indicator to tell us how much energy is being given out in its vibrations.

At first we supply equal amounts of energy to the two forks, and so hear sounds of equal loudness when they are sounded alternately. If we now continue sounding the forks alternately, but gradually increase the supply of energy to one of them, we shall notice no difference in the loudness at first; it takes some time to reach a point at which this fork sounds appreciably louder than the other. Generally speaking, we shall first notice a difference when the louder sound emits about 25 per cent. more energy than the fainter. We find in this way that our ears are insensitive to anything less than a 25 per cent. difference of energy. The pianist who is executing a rapid passage may allow himself a 25 per cent. variation in the strengths of different notes, without our ears detecting any falling off from regularity. The organ-voicer may leave a row of pipes differing by as much as 25 per cent. in strength, and even a trained ear will pass them as perfectly uniform.

If we now increase the energy supplied to both the forks —to 100 or 1000 times the original amount if we please— and repeat the experiment, we shall again find that a 25 per cent. difference is necessary before our ears are conscious of any difference at all. To produce a perceptible increase of loudness, we do not add a certain definite amount to the energy, but increase it in a certain definite proportion; it is not a matter of addition but of multiplication. We can in fact divide up the whole range of sound intensity into distinct steps, such that each is just, and only just, perceptibly louder to our ears than the preceding, and shall find that each step represents about 25 per cent. more energy than the preceding.

This illustrates a general law first enunciated by Weber: "The minimum increase of stimulus which will produce a perceptible increase of sensation is proportional to the pre-existent stimulus." The law is only approximate at best; it cannot even be tested very accurately, since it makes a statement about subjective sensations, which necessarily vary from one person to another. Even in the simple case we have just discussed, the law is known to be far from accurate. For instance, our ears are abnormally insensitive to changes of intensity in very faint sounds. When a sound has only ten times the energy it has at the threshold of hearing, twice as large a proportionate change is needed to affect our ears, as when it is reasonably loud; it is no longer a matter of 25 per cent. increase, but of 50 per cent. For still fainter sounds, and also for tones of very high or very low pitch, even larger changes are needed.

Recent experiments by Churcher, King and Davies seem to suggest that there is no range at all in which the law is entirely accurate. Moreover, while a difference of about 25 per cent. is usually needed to affect our ears, a difference of only 10 per cent. may be appreciable under the ideal conditions of an acoustical laboratory. Nevertheless, in spite of all reservations, it remains true that under normal conditions, and throughout a very large range of intensity and of pitch, the smallest increase in intensity to which our ears are sensitive is one of about 25 per cent.

To make ten such increases, we must increase the energy of the sound to $(1 \cdot 25)^{10}$ times its original intensity. The value of this number is $\frac{9765625}{1048576}$, and if we take this as equal to ten for the moment, we may say that a tenfold increase of energy—as for instance from one unit to ten units— carries us up ten steps in the scale of loudness. To get a further equal increase of loudness, we must go up another ten steps, and this requires a further tenfold increase of energy— namely from 10 to 100 units—and so on, indefinitely.

This provides an example of a very general psychological law, known as Fechner's law. It is virtually an inference from Weber's law, and states that the intensity of our sensation does not increase as rapidly as the energy of the exciting cause, but only as rapidly as the logarithm of this energy. As with the smallest perceptible change, larger changes in sensation must be reckoned in terms of multiplications, and not in terms of additions, of the exciting cause.

The Scale of Sound Intensity

The change in the intensity of a sound which results from a tenfold increase in the energy causing this sound is called a "bel". The word has nothing to do with beauty or charm, but is merely three-quarters of the surname of Graham Bell, the inventor of the telephone.

We have already thought of this tenfold increase as produced by ten equal steps of approximately 25 per cent. each. More exactly, each of these must represent an increase by a factor of $\sqrt[10]{(10)}$, of which the value is 1·2589. Each of these steps of a tenth of a bel is known as a "decibel"; as we have seen, it represents just about the smallest change in sound intensity which our ears notice under ordinary conditions.

The intensity at the threshold of hearing is usually taken as zero point, so that, if we take the smallest amount of energy we can hear as unit:

1	unit of energy gives a sound intensity of	0 decibels
1·26	,,	,,	1 decibel
1·58	,,	,,	2 decibels
2	,,	,,	3 ,,
4	,,	,,	6 ,,
8	,,	,,	9 ,,
10	,,	,,	10 ,,
100	,,	,,	20 ,,
1000	,,	,,	30 ,,

The Scale of Loudness

The scale of sound intensity had its zero fixed at the threshold of hearing, but as the position of this depends enormously on the pitch of the sound under discussion, this scale is only useful in comparing the relative loudnesses of two sounds of the same pitch. It is of no use for the comparison of two sounds of different pitches. For this latter purpose we must introduce a new scale, the scale of loudness.

The zero point of this scale is taken to be the loudness, as heard by the average normal hearer, of a sound-wave in air, which has a frequency of 1000 and a pressure range of $\frac{1}{5000}$ bar—or, more precisely, 0·0002 dyne—at the ear of the listener. This, as we have already seen, is just about the threshold of hearing for a sound of this particular frequency.

The unit on this scale is called a "phon". So long as we limit ourselves to sounds of frequency 1000, the phon is taken to be the same thing as the decibel, both as regards its amount and its zero point. Thus if a sound of frequency 1000 has an intensity of x decibels on the scale of sound intensity, it has a loudness of x phons on the new scale of loudness. But the phon and decibel diverge when the frequency of the sound is different from 1000. Two sounds of different pitch are said to have the same number of phons of loudness when they sound equally loud to the ear. Thus we say that a sound has a loudness of x phons when it sounds as loud to the ear as a sound of frequency 1000 and an intensity of x decibels. Such a sound lies at x

decibels above the threshold of hearing for a sound of frequency 1000, not above that for a sound of its own pitch.*

The Threshold of Pain

We have already considered what is the smallest amount of sound we can hear; we consider next what is the largest amount. This is not a meaningless problem. For, if we continually supply more and more energy to a source of sound—as for instance by beating a gong harder and harder—the sound will get louder and louder and, in time, we shall find it becoming too loud for pleasure. At first it is merely disagreeable, but from being disagreeable it soon passes to being uncomfortable. Finally the vibrations set up in our ear-drums and inner ear may become so violent as to give us acute pain, and possibly injure our ears.

If we note the number of bels our ears can endure without discomfort, we shall find that this again, like the position of the threshold of hearing, depends on the pitch of the sound. At the bass end of the pianoforte it is about six bels; it has risen to eleven bels by middle C; it rises further to twelve bels in the top octave of the pianoforte, after which it probably falls rapidly.

The intensity of sound at the threshold of hearing, and also the range above the threshold which we can endure without undue discomfort, both vary greatly with the pitch of the sound, but their sum, which fixes a sort of threshold of

---

* This defines the British standard phon. The Americans use the same phon as the British, but frequently describe it as a decibel. The Germans use a different zero point, 0·0003 dyne in place of 0·0002.

pain, varies much less. Throughout the greater part of the range used in music, the intensity at this threshold is given by a pressure variation of about 600 bars, except that

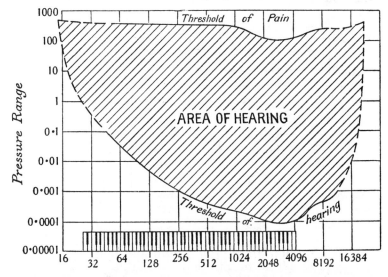

Fig. 58. The limits of the area of hearing, as determined by Fletcher and Munson. Each point in this diagram represents a sound of a certain specified frequency (as shewn on the scale at the bottom) and of a certain specified intensity (as shewn by the scale on the left). If the point lies within the shaded area, the sound can be heard with comfort. If the point lies above the shaded area, the hearing of the sound is painful. If the point lies below the shaded area, the sound lies below the threshold of hearing, and so cannot be heard at all.

it falls to about 200 bars in the region of maximum sensitivity.

We can represent this in a diagram as in fig. 58, and the shaded area which is the area of hearing can be divided up further by curves of equivalent loudness as shewn in fig. 59. Both the limits of the area of hearing and the curves of equal loudness have been determined by Fletcher and Munson.

We see at a glance how the ear is both most sensitive to faint sound, and also least tolerant to excessive sound, in

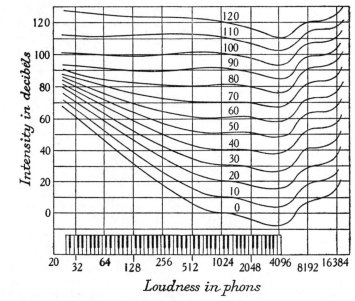

Loudness in phons

Fig. 59. The loudness of sounds which lie within the area of hearing, as determined by Fletcher and Munson. As in fig. 58, each point of the diagram represents a sound of specified frequency (as shewn on the scale at the bottom) and of specified intensity in decibels (as shewn on the scale on the left), the zero point being the faintest sound of frequency 1000 which can be heard at all. The loudness of the sound in phons is the number written on the curved line which passes through the point; thus these curves are curves of equal loudness.

the range of the upper half of the piano. To be heard at a moderate comfortable loudness of say 50 or 60 phons, treble music needs but little energy, while bass music needs a great deal. This is confirmed by exact measurements of the energy employed in playing various instruments. The following table gives the results of experiments recently made at the Bell Telephone Laboratories:

Origin of sound	Energy
	Watts
Orchestra of seventy-five performers, at loudest	70
Bass drum at loudest	25
Pipe organ at loudest	13
Trombone at loudest	6
Piano at loudest	0·4
Trumpet at loudest	0·3
Orchestra of seventy-five performers, at average	0·09
Piccolo at loudest	0·08
Clarinet at loudest	0·05
Human voice ⎰ Bass singing *ff*	0·03
Human voice ⎱ Alto singing *pp*	0·001
Average speaking voice	0·000024
Violin at softest used in a concert	0·0000038

We may notice in passing how very small is the energy of even a loud sound. A fair-sized pipe organ may need a 10,000-watt motor to blow it; of this energy only 13 watts reappears as sound, while the other 9987 watts is wasted in friction and heat. A strong man soon tires of playing a piano at its loudest, his energy output being perhaps 200 watts; of this only 0·4 watts goes into sound. A thousand basses singing *fortissimo* only give out enough energy to keep one 30-watt lamp alight; if they turned dynamos with equal vigour, 6000 such lamps could be kept alight.

The first and last entries in the table above represent the extreme range of sounds heard in a concert room, and we notice that the former is more than eighteen million times the latter. Yet this range, large though it is, is only one of 7¼ bels, and so is not much more than half of the range of 12 bels which the ear can tolerate in treble sounds.

For a person well away from the instruments, we may perhaps estimate the violin at its softest as being about 1 bel above the threshold of hearing for the note it is playing, so that the full orchestra is about 8·3 bels, or 83 decibels. This may be compared with the intensities of various other sounds, as shewn in the following table:

Threshold of hearing	0	decibels
Gentle rustle of leaves	10	,,
Quiet London garden	20	,,
Whisper at 4 feet	20	,,
Quiet suburban street, London	30	,,
Quietest time at night, Central New York	40	,,
Conversation at 12 feet	50	,,
Busy traffic, London	60	,,
Busy traffic, New York	68	,,
Very heavy traffic, New York	82	,,
Lion roaring at 18 feet	88	,,
Subway station with express passing, New York ...	95	,,
Boiler factory	98	,,
Steel plate hammered by four men, 2 feet away ...	112	,,

Owing to the different thresholds of hearing, the sounds in the above tables are not strictly comparable, unless they happen to be of the same pitch. The following table shews the differences of subjective loudness for a few common sounds:

Threshold of hearing	0	phons
Ticking of watch at 3 feet	20	,,
Sounds in a quiet residential street ...	40	,,
Quiet conversation	60	,,
Sounds in a busy main street	75	,,
Sounds in a tube train	90	,,
Sounds in a busy machine shop	100	,,
Proximity of aeroplane engine	120	,,

Experiments shew that a faint sound will not be heard at all through a louder sound of the same pitch, if the difference in intensity is more than about 1·2 bels, but the difference in loudness may be greater if the sounds are of very different pitch. Conversation at 12 feet should just be heard against busy traffic in London, because the difference in intensity only amounts to 1·0 bel; it will not, however, be heard against busy traffic in New York, because the difference here is 1·8 bels. In the same way a roaring lion would only just be heard in a boiler factory, although he might hope to attract considerable attention in a New York subway station.

Tones created by the Ear

We now pass from a discussion of the quantity of sound to a discussion of its quality. Musical sound is like light in consisting of waves of definite frequencies, and our ears are like our eyes in being sensitive only to waves whose frequencies fall within a comparatively small range. The ear is, however, endowed with a property which the eye does not possess, namely that of creating waves of entirely new frequencies out of the disturbances which fall on it. Because of this, the brain may hear tones of pitches which were entirely lacking in music as originally played.

The reason for this is comparatively easy to understand. If the ear-drum were an absolutely simple membrane like a drum-skin or a telephone diaphragm, it would vibrate in exact sympathy with any vibrations which fell upon it, at any rate so long as these were not violent enough to press it appreciably out of shape, and the vibrations which it passed on to the brain would be of precisely the same frequencies as those it received from the air. But this is only true provided that the forces tending to restore the ear-drum to its normal position, when it is forced a thousandth of an inch to the right, are just the same in amount as they are when it is forced an equal distance to the left; in other words, the ear-drum and its attachments must be supposed to form a structure with right and left symmetry. A very slight knowledge of the mechanism of the ear tells us that this is not the case; the ear-drum does not possess symmetry of this kind. On one side of it there is nothing but air; on the other side there is a complicated mechanism of bone which transmits the motion of the ear-drum to the brain. When the ear-drum moves in an outward direction, there is nothing but its own elasticity to check its motion and pull it back; when it moves inwards, its motion is further impeded by this bony structure.

Helmholtz was the first to point out that the bony structure would act in this way. Quite recently other investigators, von Békésy, and Chapin and Firestone, have found reasons for thinking that other parts of the ear—the liquid which fills the cochlea and the basilar membrane (p. 246)—produce similar results.

To trace out some of the consequences of this, let us first suppose that a simple harmonic vibration, such as the pure tone of a tuning-fork, is transmitted through the air. Simple harmonic pressure-waves now fall on the ear-drum, and if this possessed perfect fore-and-aft symmetry, its response would be a simple harmonic vibration, such as that shewn by the thick line in fig. 60. Actually the presence of the bony mechanism impedes the motion of the ear-drum in one direction—say that which is represented below the horizontal line in fig. 60—with the result that here the response of the ear-drum will be restricted, and its displacement will be more like that shewn by the dotted line.

To discover what tones are conveyed to the brain, we

must analyse the amended displacement curve into its
constituent simple harmonic curves. The new curve is
represented by the thick curve above the line and by the
dotted line below. If the original sound vibration had a
frequency of 200 a second, this curve must clearly repeat
itself 200 times a second, and Fourier's theorem (p. 79)
tells us that its constituent simple harmonic curves will
be of frequencies 200, 400, 600, 800, etc. In other words
the want of symmetry of the ear-drum has added new
frequencies 400, 600, 800, etc. to the original frequency
of 200. In terms of music, the ear-drum not only transmits

Fig. 60. The response of the ear-drum to a simple sound. The unsymmetrical
nature of the human ear results in parts of the simple harmonic curve,
which is drawn thick, being replaced by those shewn by broken lines.

the tone which originally fell on it, but adds the octave
and all the other natural harmonics.

Let us next suppose that vibrations of frequencies 200
and 300 fall simultaneously on the ear-drum. If the ear
were a symmetrical structure, its response to the two
vibrations separately would be two simple harmonic curves
of frequencies 200 and 300 performed simultaneously—
as represented by the thin lines in fig. 61. The total response would be represented by the superposition of these
two curves, which is shewn by the thick line. But the
want of symmetry in the ear-drum again results in the
replacement of part of this curve by another such as is
shewn by the dotted line. And if we wish to know what
tones the ear-drum passes on to the brain, we must again

resolve this amended curve into its constituent simple harmonic curves. Since the curve repeats itself with a frequency of 100, the constituent curves will have frequencies of 100, 200, 300, 400, 500, 600, etc. The frequencies 200 and 300 represent the original tones, while the frequencies 400 and 600 represent their octaves, which we have already seen must be present. The remaining frequencies 100 and 500 represent entirely new tones. The former is described as a "difference tone" because its frequency, 100, is the difference of two original frequencies;

Fig. 61. The response of the ear-drum to two simple tones sounded simultaneously. As in fig. 60, a part of the thick curve is replaced by that shewn by the broken line.

the latter is described as a "summation tone", its frequency, 500, being the sum of the original frequencies.

If the two original tones had been of frequencies 200 and 201, the problem could have been treated in the same way. The displacement curve would now only repeat itself once a second, so that when it was analysed by Fourier's theorem, the constituent curves would be found to have frequencies 1, 2, 3, 4, etc. There is no means of telling, from general principles alone, which of this vast array of frequencies represent audible sounds, and which are unimportant. The question can only be settled by a detailed, and somewhat complicated, mathematical investigation.

Such an investigation was first made by Helmholtz. He supposed the ear to be unsymmetrical in the way we have already explained. On this basis he shewed, in the first place, that a single pure tone, conveyed to the ear-drum by simple harmonic vibrations of the surrounding air, will not produce merely a simple harmonic vibration of the same frequency in the ear-drum. In addition to this, superposed on to it by the asymmetry of the ear, there will be a second simple harmonic vibration of just double the frequency. In other words the ear of its own accord, and by means of its asymmetry, brightens up the pure tone by adding something of its octave.

He further shewed that when two pure tones are sounded simultaneously, the ear of its own accord not only adds their two octaves, but also summation and difference tones having the frequencies already explained.

Finally, when a whole lot of pure tones of frequencies p, q, r, etc. are sounded simultaneously the ear not only adds their octaves, of frequencies $2p$, $2q$, $2r$, but also all their summation tones, of frequencies $p+q$, $q+r$, $p+r$, and all their difference tones of frequencies $p-q$, $q-r$, $p-r$.

Helmholtz supposed that the asymmetry of the ear had only a slight influence in modifying the tones, so that the octaves and difference tones were very faint. This is true when the fundamental tones are themselves quite faint, but not when they are loud. If tones of frequencies p, q and r are sounded loudly and simultaneously, it can be shewn that we shall hear tones of the frequencies shewn in the table overleaf.

Loudest of all $\left.\begin{array}{c} p \\ q \\ r \end{array}\right\}$ the fundamental tones.

Next loudest $\left.\begin{array}{c} 2p \\ 2q \\ 2r \end{array}\right\}$ the second harmonics of the foregoing.

$\left.\begin{array}{c} p+q \\ q+r \\ p+r \end{array}\right\}$ the first summation tones.

$\left.\begin{array}{c} p-q \\ q-r \\ p-r \end{array}\right\}$ the first difference tones.

Next loudest $\left.\begin{array}{c} 3p \\ 3q \\ 3r \end{array}\right\}$ the third harmonics.

$\left.\begin{array}{c} p+q+r \\ 2p+q \\ 2q+p \\ 2q+r \\ 2r+q \\ 2r+p \\ 2p+r \end{array}\right\}$ the second summation tones.

$\left.\begin{array}{c} 2p-q \\ p-r-q, \text{ etc.} \end{array}\right\}$ the second difference tones.

To take a concrete instance, let us suppose that the three original frequencies p, q, r are those of the notes c', e', g', the fourth, fifth and sixth harmonics of CC. Then the full table will stand as follows:

The fundamental tones	Harmonics 4, 5, 6 of CC.
The first difference tones	CC (twice) and harmonic 2 of CC.
The second harmonics	Harmonics 8, 10, 12 of CC.
The first summation tones	Harmonics 9, 10, 11 of CC.
The second difference tones	Harmonics 2, 3, 4, 5, 6, 7, 8 of CC.
The third harmonics	Harmonics 12, 15, 18 of CC.
The second summation tones	Harmonics 13, 14, 15, 16, 17 of CC.

Thus when the fourth, fifth and sixth harmonics are sounded without the fundamental tone, the ear adds the fundamental and all harmonics up to the eighteenth.

It is in general true that when any two or more pure tones which are sounded simultaneously happen to be harmonics of the same fundamental note, then the ear adds this fundamental note and many of its harmonics, of its own accord—a result of tremendous importance in all branches of pure and applied acoustics. If the pure tones are all the odd-numbered harmonics of a fundamental note, then the ear of its own accord adds all the even harmonics. If the two pure tones differ only slightly in frequency, then their "difference tone" has the same frequency as the beats we have already discussed (p. 46), so that as the two original tones approximate to one another, their difference tone degenerates into beats, while their summation tone approaches to their second harmonic.

Difference and Summation Tones

Difference tones were discovered by a German organist Sorge in 1745 and again independently by the famous violinist Tartini in 1754. They can be heard quite easily on loudly sounding almost any two notes which are a fifth apart; the tone an octave below the lower note is then heard. For instance we may play c' and g' forcibly on the piano, and shall hear C. They may also be heard when voices, and particularly treble voices, are singing in harmony, the difference tone providing a dim bass accompaniment (see fig. 62, below). For instance, if sopranos sing c", e", we hear the difference tone C as a bass. If c", e" are sung in just intonation, the C is in perfect tune with them, but if c″, e″ are sung in equal temperament, the difference tone lies nearer to C♯ than to C, providing what Helmholtz describes as "a horrible bass, which is all the more annoying for coming tolerably near to the correct bass". Indeed Helmholtz considered that these dissonant combination tones formed "the most annoying part of equally tempered harmonies".

Summation tones were discovered by Helmholtz in 1856. They are much more difficult to hear than difference tones —largely because they lie in a region of frequencies which is already occupied by the harmonics of the original sounds.

There has never been any doubt as to these various tones being heard, but there has been a good deal of discussion as to whether they are purely subjective or not. If they were heard only as a consequence of the asymmetry of the eardrum, then they would of course only exist inside the ear, and could not be picked up by any resonator outside the ear. After it had been believed for some time that the notes were purely subjective in this sense, Helmholtz was able to prove that they existed objectively, and although his experiments have repeatedly been challenged, no one any longer doubts that under certain conditions, the tones have an objective existence and can be picked up with a resonator. Indeed we can easily prove this for ourselves by striking c′ and g′ with sufficient force on the piano. We not only hear C, but shall also find that the C strings are feebly vibrating, as can be verified by laying small chips of wood across them, in the way explained on p. 58.

The condition that these tones shall have an objective existence is very simple. All sound is represented by a curve, and this curve undergoes a certain amount of distortion each time the sound passes from one conveyor to another. If any one of the conveyors is unsymmetrical in the same way as the ear-drum, then this asymmetry must have the same kind of effect on the sound-curve as asymmetry of the ear-drum has, and must necessarily produce sum and difference tones in precisely the way already

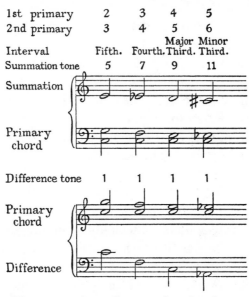

1st primary	2	3	4	5
2nd primary	3	4	5	6
			Major	Minor
Interval	Fifth.	Fourth.	Third.	Third.
Summation tone	5	7	9	11

Summation

Primary chord

| Difference tone | 1 | 1 | 1 | 1 |

Primary chord

Difference

Fig. 62. The difference tone is usually concordant, but the summation tone discordant, with the tones which produce it.

explained. We may properly think of these tones as impurities which necessarily become mixed with the pure tones whenever the sound-curve is transmitted by any unsymmetrical structure whatever. And a little thought will shew that the majority of transmitting structures are unsymmetrical, in a greater or less degree.

Theory and experience agree in shewing that when the

primary tones are sounded faintly, both the difference and the summation tones are heard very faintly indeed, the summation tones especially so. This is fortunate for our enjoyment of music, since tones which are perfectly consonant in themselves may quite well produce dissonant summation tones. This is shewn in fig. 62. On the other hand, as the last line shews, the difference tones, which happily sound louder than the summation tones, are usually concordant.

While quite properly thinking of summation and difference tones as impurities, we must not forget that impurities may be either harmful or beneficent. On the whole, summation tones must be placed in the category of harmful impurities, but difference tones can be turned to advantage in a variety of ways.

Practical Uses of Difference Tones

One of the most striking instances of this is provided by the ordinary telephone. A telephone diaphragm, like every other structure, has its own free vibrations, and the frequencies of these happen to lie mostly within the range of frequencies covered by the human voice. There is, then, a danger that the diaphragm may over-emphasise, through resonance, those particular notes of the human voice which happen to coincide with its own free periods, while leaving the others inaudible. A poor telephone often does this, transforming the voice into a succession of metallic claps of sound. A good modern telephone is so designed that the free vibrations of its diaphragm are spread as uniformly as possible over a range of frequencies from about 300 to 2400, so that sounds within this range are transmitted in considerable strength, while those above and below it are hardly transmitted at all. Now the main frequencies of both male and female voices lie below this range, so that the telephone transmits very little of the main tones of a conversation. It transmits chiefly harmonics, and out of these the ear-drum of the listener reconstructs the main tones as difference tones, which are then transmitted to our brains in considerable strength. The Bell Telephone Company of America has constructed two sets of gramophone records which shew this very clearly. On the one set we hear a singing voice, a speaking voice, organ tone and so forth, all faithfully reproduced. On the other we hear the same voices and instruments with the fundamental tones deliberately cut out. Yet the second set of records sounds almost exactly like the first, the speech and music being perfectly intelligible although all fundamental tones are missing, except in the form of difference tones created by the ear of the listener.

We find much the same thing happening in the loudspeakers of our radio sets. Many are designed deliberately to cut out all frequencies below about 250, the frequency of about middle C, and so transmit no bass or tenor tones at all. Yet we hear the double bass strings, the basses of the brass, and male voices with absolute clearness. The explanation is, of course, that all these sources of sound are rich in harmonics. Out of these our ears create the missing fundamental tones and lower harmonics as difference tones, and the combination of these with the higher harmonics, which come through unhindered, restores for us the tone played by the orchestra. Obviously this can only happen if harmonics are transmitted in abundance. A deep-pitched tuning-fork would not be heard at all, because it has no harmonics. The drum does not come through in its proper pitch, because the frequencies of its free vibrations do not form a series of natural harmonics (i.e. tones with frequencies in the ratio $1 : 2 : 3 : 4 : 5 :...$). The same is true of cymbals, of many bells and of percussion instruments in general, and incidentally of the applause at the end of a concert item—this is mere noise without harmonic overtones, so that all its deeper constituents are missing.

Difference tones are also turned to advantage in a type of whistle which is used by football referees and the police. When this is blown, a blast of air is distributed equally over the mouths of two pipes of slightly unequal length. These of course sound two notes of slightly different frequencies, and we hear both these and their difference tone, which we can make as deep as we please by making the pipes nearly equal to one another in length. For instance, pipes of lengths 2 and $2\frac{1}{8}$ inches give a difference tone whose pitch is that of a 3-foot pipe. The pipes may even be so short that their individual tones are above the range of audition. Then neither pipe can be heard when sounded separately, but when the two pipes are sounded together, the difference to e is heard quite clearly.

There is no limit to the depth of tone that can be obtained in this way, except that fixed by the limitations of the human ear; as we make the pipes more nearly equal in length, the tone finally becomes so deep that the ear is insensitive to it. If we still go on making the two tones more nearly equal, the difference tone reappears in the form of beats (p. 237).

The organ-builder utilises this principle to get tones of

low pitch without incurring the expense of pipes of great length. For example, the longest pipe in a big organ, CCCC, is about 32 feet long, and its fundamental tone has a frequency of 16, while its harmonics have frequencies of 32, 48, 64, and so on. The fundamental note is inaudible to most people, except perhaps for a small vibration transmitted through floors and walls, which is felt rather than heard. What our ears hear consists mainly of harmonics of frequencies 32, 48, and so on, sounding the notes CCC, GG, etc. Thus this immense and expensive pipe is serving no purpose beyond producing tones of 16 and $10\frac{2}{3}$ pitch as harmonics for our ears to recombine into difference tones, and these might equally well be produced by smaller and cheaper pipes. For this reason the largest pipes in an organ are often replaced by pairs of shorter pipes sounding the second and third harmonics respectively of the desired tone. The combination is known as an "acoustic bass"; if well designed, it may give the impression that we are hearing a single CCCC pipe, although we cannot escape hearing its third harmonic also in quite disproportionate strength.

Much the same thing appears to happen in the lowest notes of the piano. The fundamental tones of these are so deep that the ear only hears them when they are sounded with immense energy, in which case the upper harmonics sound intolerably loud. Under ordinary conditions, it seems likely that the ear hears only the harmonics directly, and the fundamental note only as a difference tone.

At the other end of the musical scale the same principle is utilised in the "mixture" stops of the organ. This is a generic name applied to a number of stops (mixture,

cornet, cymbal, sesquialtera, furniture, harmonics, etc.) in
which each key sounds several small pipes of high pitch.
Usually, but not always, the notes sounded by the separate
pipes are harmonics of the fundamental note. For instance
if we sound bottom C of the organ, a mixture stop may
sound any of the following notes:

Fig. 63.

The modern mixture stop is used mainly to add brilliance
to the tone. As compared with the tone of the piano (p. 91)
or violin (p. 76), or the wind instruments of the orchestra
(p. 147), organ tone is deficient in those higher harmonics
which make for brilliance, so that brilliance can only be
attained by introducing these harmonics artificially in the
form of sound from separate pipes. But in old organs
the mixture stop often served an entirely different purpose, the harmonics provided by its shrill pipes combining
to produce the fundamental as a combination tone, much
as in the "acoustic bass" of to-day. Some of the old organs
which are still to be found in North Germany, Holland and
Northern Italy would seem from their specifications to be
a miscellaneous collection of mixture ranks, but when
heard are found to produce a deep rich tone from a series
of very small pipes. By using pipes of the pitches to which
the ear is most sensitive, the organ-builders of past days
could fill a church with rich full organ tone out of an

instrument which occupied less room than an upright piano, and needed amazingly little wind to blow it. On the other hand, this plan had its disadvantages. As each individual pipe of the mixture gave out its own harmonics in addition to its fundamental tone, the effect of too many ranks of mixtures was apt to be shrill and discordant, and whenever possible the purer tones of larger pipes were introduced as well. For instance the famous Schnitger organ in S. Jacobi at Hamburg has eleven ranks of mixtures on the pedals, but it has two 32-foot registers as well; the manuals contain thirty-nine ranks of mixtures and mutations, but also four registers of 16-foot pitch.

The Mechanism of the Ear

It would form a pleasing and perfect ending to our book if we could explain how the ear comes to have all the remarkable capacities we have noted. Unhappily this is still beyond the powers of science; no completely satisfactory theory of hearing, or of the mechanism of the ear, has yet been found.

So far our explorations have taken us no farther into the ear than the ear-drum and the chain of small bones, the ossicles, which lie immediately behind it. This chain of bones transmits the motion of the ear-drum to a second membrane—the "oval" membrane of the cochlea (fig. 1)—and incidentally serves as a safety device acting rather like the slipping clutch of a motor-car, which may save the oval membrane from injury, if the ear-drum is set too violently into sudden motion. On passing through the oval membrane, we find ourselves inside the hard bony structure of the cochlea. Viewed from outside this looks somewhat like a snail's shell, or a coiled tube. The tube has a total length of about $1\frac{1}{4}$ inches; its diameter decreases steadily as we pass along its length, the average being only about $\frac{1}{16}$ inch. The tube is hollow inside, and is divided longitudinally into two approximately equal parts, so that each has a length of $1\frac{1}{4}$ inch, an average width of $\frac{1}{16}$ inch, and an average height of $\frac{1}{32}$ inch. These two divisions may be described as the upper and lower galleries. Both are filled with fluid and the only connection between them is a small opening, known as the helicotrema, at the narrow

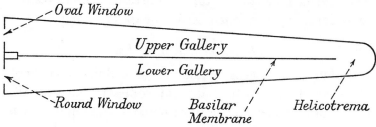

Fig. 64. Diagrammatic representation of the cochlea, uncoiled.

end. The oval membrane itself forms a sort of window at the broad end of the upper gallery, while a similar but circular membrane forms a window to the lower gallery. Apart from the coiling of the tube of the cochlea, the arrangement is that shewn diagrammatically in fig. 64.

When the ear-drum is set into vibration, its motion is transmitted through the ossicles to the oval membrane, and from this to the fluid in the cochlea. Waves of pressure then travel through the fluid in the upper gallery, pass through the helicotrema, along the lower gallery, and finally expend their energy in producing vibrations in the circular membrane.

All the mechanism so far described has served merely to send these pressure-waves along the two galleries of the cochlea, and prevent their being reflected back again. If the floor separating these galleries were an ordinary floor, this would be the end of the matter, but it is no ordinary floor. Over about half its width, it consists of bone—a sort of balcony projecting out from the wall of the cochlea. The other, and the more remarkable, half consists of a thin continuous membrane, the basilar membrane, the structure of which is strengthened by an immense number of tightly stretched fibres. If we have to compare it with a familiar object, by far the best to choose is a grand piano, built with enormous complexity, but on a diminutive scale. Although the membrane is only an eight-thousandth of an inch in thickness, and $1\frac{1}{4}$ inches in length, yet about 24,000 fibres are embedded in it, ranging in length from a fifteenth to a 170th of an inch. The short fibres which correspond to the treble wires of the piano are very tightly stretched, while those at the other end—the bass wires—are much looser.

The whole structure is on so small a scale that there is a temptation to dismiss the comparison with a piano as fanciful. There is, however, very convincing evidence that the various "strings" of the instrument are associated with notes of different pitch. Experimenters have been able to damage selected bits of the basilar membranes of animals—as, for instance, by drilling minute holes through the wall of the cochlea—and find that the animal then becomes deaf to notes of certain pitches and no others; in this way it is possible to map out the basilar membrane in terms of the frequencies of sound of different pitches.

The converse experiment can also be performed. An animal is made deaf to a note of any selected pitch, and the corresponding section of its basilar membrane is then found to have been damaged.

This makes it abundantly clear that each fibre is in some way associated with a definite frequency of sound, and it is natural to suspect that these frequencies are those of the free vibrations of the fibres. It may seem incredible that a fibre less than a fifteenth of an inch in length can be made to emit the same note as a piano wire 6 feet in length—for this is implied in the hypothesis we are considering—but calculation suggests that there is no impossibility. We can lower the pitch of a stretched string by adding to its mass—as indeed is commonly done in the piano by coiling copper wire round the bass strings. The fibres of the basilar membrane are embedded in the membrane itself and, as this shares the motion of the fibres, it adds to their effective mass. Moreover the motion of the membrane sets the fluid which fills the cochlea into motion, so that this also adds to the effective mass of the fibres, especially at the bass end, where the galleries are widest.

Beatty has estimated that if the bass fibres are stretched to a tension of about 3 lb. to the square inch of cross-section, they will vibrate in the frequencies required of them. He finds also that for the treble fibres to vibrate in their required frequencies, they must be stretched to a tension of about 4 tons to the square inch. This is a large, but by no means impossible, tension. Human hair can be stretched to a tension of about 9 tons to the square inch before breaking, while catgut and silkworm gut will survive tensions of 27 and 32 tons to the square inch respectively. A steel pianoforte wire does not break until the tension reaches about 150 tons to the square inch.

Thus there is every reason to think that the basilar membrane and the fibres embedded in it have most of the properties of the sound-board and wires of a grand pianoforte. We have seen how the latter may be treated as a series of resonators; sound-waves passing through the air set the sound-board into vibration, and the wires of the same frequency as these sound-waves are then set into strong vibration by resonance. In the same way sound-waves passing through the fluid of the cochlea must set the basilar membrane into vibration, and probably cause the fibres of their own frequency to vibrate strongly. It is now easy to imagine a separate nerve connecting each fibre to the brain, and minute currents in these nerves keeping the brain informed as to the vibrations of the different fibres, and so of the composition of any sound which may be falling on the ear-drum.

Such, in broad outline, is the "resonance theory of hearing". It explains at once why the ear can resolve a chord into its constituent notes of different pitches. The eye does not possess this power, for it cannot resolve a blend of colours into the separate pure colours of which it is constituted, and this alone shews that the mechanism of the ear must be very different from that of the eye. The resonance theory also accounts for the law of Ohm—that the ear cannot distinguish phase differences in the constituent vibrations out of which a composite tone is formed. In brief, it does not inform the brain as to the mathematical shape of the sound-curve it is hearing, but only as to the way in which the energy is distributed over the different pure tones. No theory of hearing which fails to explain these two outstanding facts can be seriously considered; the resonance theory passes the two tests triumphantly. Many other theories of hearing have been propounded, comparing the mechanism of the ear to that of a telephone, a flexible diaphragm, a pulsating artery, and so forth, but most fail under these tests.

On the other hand, when attempts are made to develop the resonance theory in greater detail than has been described in the present book, difficulties are encountered at once.

The galleries above and below the basilar membrane are not empty, and could not be, since they contain the machinery needed to transform the purely mechanical vibrations of the fibre into electric currents, as well as the nerves which carry these currents to the brain. Within the small volume of the cochlea lie 24,000 minute electrical machines, and from the central axis of the cochlea emerges a cable of 24,000 conducting nerves. This machinery is inevitably complicated—so complicated that science has not yet completely discovered its workings.

Hearing in Animals

If the ear of man is surprising in its intricacy and powers, those of the lower animals are even more so. The ears of mammals are generally similar to those of man, but simpler. The cat, for instance, has 3 complete turns in its cochlea, as against man's $2\frac{3}{4}$, but its basilar membrane has only 16,000 fibres as against the 24,000 of man. Reptiles and birds have somewhat similar but again simpler ears, birds having only about 3000 fibres in their basilar membrane.

For a long time insects were thought to have entirely different organs of hearing, but recent knowledge shews that they have ears which are in many respects similar to those of the vertebrates, although they are often located in what seem strange places to us, such as the abdomen, thorax, or even on the forelegs, just below the knee-joint.

On the ordinary grasshopper, the ear-drums can be seen as little circular membranes—one on each side of the lower part of the body—and behind these lie inner ears which shew many resemblances to those of man. There are other grasshoppers whose ears are in their forelegs. Professor Regen of Vienna has specially studied the ears and hearing of the grasshopper *Thamnotrizon Apterus*. The females are courted in a sort of singing, or rather chirping, contest; he who chirps best wins the prize. There can be no question that the "sex-appeal" operates through the sense of hearing, and not, for instance, through sight or smell, since the female can be excited by telephone; if the male can be persuaded to chirp at one end of a telephone line, a female at the other end will jump into the air and settle down to listen in front of the receiver. The male can also be sent up in a small balloon, when the females follow to listen to him. Under more normal conditions the female flies directly towards that male which attracts her most. But if the membranes just below her knee-joints are punctured or destroyed, she can no longer hear anything, and no such flight takes place; if only one of the two membranes is destroyed the female still seeks her male, but her sense of his direction is lost; she flies in a lopsided way, and only finds him after a succession of misdirected flights.

To mention only one other instance, various kinds of

butterflies and moths have a pair of ears on the thorax. Their range of hearing is small, being generally limited to tones of high pitch. The ordinary notes of a musical instrument have no effect on them, but a high screech or squeak, such as is made by drawing a piece of cork over glass or turning a glass stopper round in the neck of a bottle, will make them change the direction of their flight. If, however, the membranes on the thorax have been destroyed, no such reaction takes place, shewing that their sense of hearing is really located in their waists.*

Students of evolution in the animal world tell us that the ear was the last of the sense-organs to arrive; it is beyond question the most intricate and the most wonderful.

* Many other interesting examples are to be found in Beatty's *Hearing in Man and Animals* (Bell, 1932), from which much of the above information is taken.

INDEX

A CATALOG OF SELECTED
DOVER BOOKS
IN SCIENCE AND MATHEMATICS

QUALITATIVE THEORY OF DIFFERENTIAL EQUATIONS, V.V. Nemytskii and V.V. Stepanov. Classic graduate-level text by two prominent Soviet mathematicians covers classical differential equations as well as topological dynamics and ergodic theory. Bibliographies. 523pp. 5⅜ × 8½. 65954-2 Pa. $14.95

MATRICES AND LINEAR ALGEBRA, Hans Schneider and George Phillip Barker. Basic textbook covers theory of matrices and its applications to systems of linear equations and related topics such as determinants, eigenvalues and differential equations. Numerous exercises. 432pp. 5⅜ × 8½. 66014-1 Pa. $10.95

QUANTUM THEORY, David Bohm. This advanced undergraduate-level text presents the quantum theory in terms of qualitative and imaginative concepts, followed by specific applications worked out in mathematical detail. Preface. Index. 655pp. 5⅜ × 8½. 65969-0 Pa. $14.95

ATOMIC PHYSICS (8th edition), Max Born. Nobel laureate's lucid treatment of kinetic theory of gases, elementary particles, nuclear atom, wave-corpuscles, atomic structure and spectral lines, much more. Over 40 appendices, bibliography. 495pp. 5⅜ × 8½. 65984-4 Pa. $12.95

ELECTRONIC STRUCTURE AND THE PROPERTIES OF SOLIDS: The Physics of the Chemical Bond, Walter A. Harrison. Innovative text offers basic understanding of the electronic structure of covalent and ionic solids, simple metals, transition metals and their compounds. Problems. 1980 edition. 582pp. 6⅛ × 9¼. 66021-4 Pa. $16.95

BOUNDARY VALUE PROBLEMS OF HEAT CONDUCTION, M. Necati Özisik. Systematic, comprehensive treatment of modern mathematical methods of solving problems in heat conduction and diffusion. Numerous examples and problems. Selected references. Appendices. 505pp. 5⅜ × 8½. 65990-9 Pa. $12.95

A SHORT HISTORY OF CHEMISTRY (3rd edition), J.R. Partington. Classic exposition explores origins of chemistry, alchemy, early medical chemistry, nature of atmosphere, theory of valency, laws and structure of atomic theory, much more. 428pp. 5⅜ × 8½. (Available in U.S. only) 65977-1 Pa. $11.95

A HISTORY OF ASTRONOMY, A. Pannekoek. Well-balanced, carefully reasoned study covers such topics as Ptolemaic theory, work of Copernicus, Kepler, Newton, Eddington's work on stars, much more. Illustrated. References. 521pp. 5⅜ × 8½. 65994-1 Pa. $12.95

PRINCIPLES OF METEOROLOGICAL ANALYSIS, Walter J. Saucier. Highly respected, abundantly illustrated classic reviews atmospheric variables, hydrostatics, static stability, various analyses (scalar, cross-section, isobaric, isentropic, more). For intermediate meteorology students. 454pp. 6⅛ × 9¼. 65979-8 Pa. $14.95

RELATIVITY, THERMODYNAMICS AND COSMOLOGY, Richard C. Tolman. Landmark study extends thermodynamics to special, general relativity; also applications of relativistic mechanics, thermodynamics to cosmological models. 501pp. 5⅜ × 8½.   65383-8 Pa. $13.95

APPLIED ANALYSIS, Cornelius Lanczos. Classic work on analysis and design of finite processes for approximating solution of analytical problems. Algebraic equations, matrices, harmonic analysis, quadrature methods, much more. 559pp. 5⅜ × 8½.   65656-X Pa. $13.95

INTRODUCTION TO ANALYSIS, Maxwell Rosenlicht. Unusually clear, accessible coverage of set theory, real number system, metric spaces, continuous functions, Riemann integration, multiple integrals, more. Wide range of problems. Undergraduate level. Bibliography. 254pp. 5⅜ × 8½.   65038-3 Pa. $8.95

INTRODUCTION TO QUANTUM MECHANICS With Applications to Chemistry, Linus Pauling & E. Bright Wilson, Jr. Classic undergraduate text by Nobel Prize winner applies quantum mechanics to chemical and physical problems. Numerous tables and figures enhance the text. Chapter bibliographies. Appendices. Index. 468pp. 5⅜ × 8½.   64871-0 Pa. $12.95

ASYMPTOTIC EXPANSIONS OF INTEGRALS, Norman Bleistein & Richard A. Handelsman. Best introduction to important field with applications in a variety of scientific disciplines. New preface. Problems. Diagrams. Tables. Bibliography. Index. 448pp. 5⅜ × 8½.   65082-0 Pa. $12.95

MATHEMATICS APPLIED TO CONTINUUM MECHANICS, Lee A. Segel. Analyzes models of fluid flow and solid deformation. For upper-level math, science and engineering students. 608pp. 5⅜ × 8½.   65369-2 Pa. $14.95

ELEMENTS OF REAL ANALYSIS, David A. Sprecher. Classic text covers fundamental concepts, real number system, point sets, functions of a real variable, Fourier series, much more. Over 500 exercises. 352pp. 5⅜ × 8½. 65385-4 Pa. $11.95

PHYSICAL PRINCIPLES OF THE QUANTUM THEORY, Werner Heisenberg. Nóbel Laureate discusses quantum theory, uncertainty, wave mechanics, work of Dirac, Schroedinger, Compton, Wilson, Einstein, etc. 184pp. 5⅜ × 8½.   60113-7 Pa. $6.95

INTRODUCTORY REAL ANALYSIS, A.N. Kolmogorov, S.V. Fomin. Translated by Richard A. Silverman. Self-contained, evenly paced introduction to real and functional analysis. Some 350 problems. 403pp. 5⅜ × 8½.   61226-0 Pa. $10.95

PROBLEMS AND SOLUTIONS IN QUANTUM CHEMISTRY AND PHYSICS, Charles S. Johnson, Jr. and Lee G. Pedersen. Unusually varied problems, detailed solutions in coverage of quantum mechanics, wave mechanics, angular momentum, molecular spectroscopy, scattering theory, more. 280 problems plus 139 supplementary exercises. 430pp. 6½ × 9¼.   65236-X Pa. $13.95

ASYMPTOTIC METHODS IN ANALYSIS, N.G. de Bruijn. An inexpensive, comprehensive guide to asymptotic methods—the pioneering work that teaches by explaining worked examples in detail. Index. 224pp. 5⅜ × 8½.   64221-6 Pa. $7.95

OPTICAL RESONANCE AND TWO-LEVEL ATOMS, L. Allen and J.H. Eberly. Clear, comprehensive introduction to basic principles behind all quantum optical resonance phenomena. 53 illustrations. Preface. Index. 256pp. 5⅜ × 8½.
65533-4 Pa. $8.95

COMPLEX VARIABLES, Francis J. Flanigan. Unusual approach, delaying complex algebra till harmonic functions have been analyzed from real variable viewpoint. Includes problems with answers. 364pp. 5⅜ × 8½.   . 61388-7 Pa. $9.95

ATOMIC SPECTRA AND ATOMIC STRUCTURE, Gerhard Herzberg. One of best introductions; especially for specialist in other fields. Treatment is physical rather than mathematical. 80 illustrations. 257pp. 5⅜ × 8½.   60115-3 Pa. $6.95

APPLIED COMPLEX VARIABLES, John W. Dettman. Step-by-step coverage of fundamentals of analytic function theory—plus lucid exposition of five important applications: Potential Theory; Ordinary Differential Equations; Fourier Transforms; Laplace Transforms; Asymptotic Expansions. 66 figures. Exercises at chapter ends. 512pp. 5⅜ × 8½.   64670-X Pa. $12.95

ULTRASONIC ABSORPTION: An Introduction to the Theory of Sound Absorption and Dispersion in Gases, Liquids and Solids, A.B. Bhatia. Standard reference in the field provides a clear, systematically organized introductory review of fundamental concepts for advanced graduate students, research workers. Numerous diagrams. Bibliography. 440pp. 5⅜ × 8½.   64917-2 Pa. $11.95

UNBOUNDED LINEAR OPERATORS: Theory and Applications, Seymour Goldberg. Classic presents systematic treatment of the theory of unbounded linear operators in normed linear spaces with applications to differential equations. Bibliography. 199pp. 5⅜ × 8½.   64830-3 Pa. $7.95

LIGHT SCATTERING BY SMALL PARTICLES, H.C. van de Hulst. Comprehensive treatment including full range of useful approximation methods for researchers in chemistry, meteorology and astronomy. 44 illustrations. 470pp. 5⅜ × 8½.   64228-3 Pa. $11.95

CONFORMAL MAPPING ON RIEMANN SURFACES, Harvey Cohn. Lucid, insightful book presents ideal coverage of subject. 334 exercises make book perfect for self-study. 55 figures. 352pp. 5⅜ × 8¼.   64025-6 Pa. $11.95

OPTICKS, Sir Isaac Newton. Newton's own experiments with spectroscopy, colors, lenses, reflection, refraction, etc., in language the layman can follow. Foreword by Albert Einstein. 532pp. 5⅜ × 8½.   60205-2 Pa. $11.95

GENERALIZED INTEGRAL TRANSFORMATIONS, A.H. Zemanian. Graduate-level study of recent generalizations of the Laplace, Mellin, Hankel, K. Weierstrass, convolution and other simple transformations. Bibliography. 320pp. 5⅜ × 8½.   65375-7 Pa. $8.95

THE ELECTROMAGNETIC FIELD, Albert Shadowitz. Comprehensive undergraduate text covers basics of electric and magnetic fields, builds up to electromagnetic theory. Also related topics, including relativity. Over 900 problems. 768pp. 5⅜ × 8¼. 65660-8 Pa. $18.95

FOURIER SERIES, Georgi P. Tolstov. Translated by Richard A. Silverman. A valuable addition to the literature on the subject, moving clearly from subject to subject and theorem to theorem. 107 problems, answers. 336pp. 5⅜ × 8½. 63317-9 Pa. $9.95

THEORY OF ELECTROMAGNETIC WAVE PROPAGATION, Charles Herach Papas. Graduate-level study discusses the Maxwell field equations, radiation from wire antennas, the Doppler effect and more. xiii + 244pp. 5⅜ × 8½. 65678-0 Pa. $6.95

DISTRIBUTION THEORY AND TRANSFORM ANALYSIS: An Introduction to Generalized Functions, with Applications, A.H. Zemanian. Provides basics of distribution theory, describes generalized Fourier and Laplace transformations. Numerous problems. 384pp. 5⅜ × 8½. 65479-6 Pa. $11.95

THE PHYSICS OF WAVES, William C. Elmore and Mark A. Heald. Unique overview of classical wave theory. Acoustics, optics, electromagnetic radiation, more. Ideal as classroom text or for self-study. Problems. 477pp. 5⅜ × 8½. 64926-1 Pa. $12.95

CALCULUS OF VARIATIONS WITH APPLICATIONS, George M. Ewing. Applications-oriented introduction to variational theory develops insight and promotes understanding of specialized books, research papers. Suitable for advanced undergraduate/graduate students as primary, supplementary text. 352pp. 5⅜ × 8½. 64856-7 Pa. $9.95

A TREATISE ON ELECTRICITY AND MAGNETISM, James Clerk Maxwell. Important foundation work of modern physics. Brings to final form Maxwell's theory of electromagnetism and rigorously derives his general equations of field theory. 1,084pp. 5⅜ × 8½. 60636-8, 60637-6 Pa., Two-vol. set $23.90

AN INTRODUCTION TO THE CALCULUS OF VARIATIONS, Charles Fox. Graduate-level text covers variations of an integral, isoperimetrical problems, least action, special relativity, approximations, more. References. 279pp. 5⅜ × 8½. 65499-0 Pa. $8.95

HYDRODYNAMIC AND HYDROMAGNETIC STABILITY, S. Chandrasekhar. Lucid examination of the Rayleigh-Benard problem; clear coverage of the theory of instabilities causing convection. 704pp. 5⅜ × 8¼. 64071-X Pa. $14.95

CALCULUS OF VARIATIONS, Robert Weinstock. Basic introduction covering isoperimetric problems, theory of elasticity, quantum mechanics, electrostatics, etc. Exercises throughout. 326pp. 5⅜ × 8½. 63069-2 Pa. $8.95

DYNAMICS OF FLUIDS IN POROUS MEDIA, Jacob Bear. For advanced students of ground water hydrology, soil mechanics and physics, drainage and irrigation engineering and more. 335 illustrations. Exercises, with answers. 784pp. 6⅛ × 9¼. 65675-6 Pa. $19.95

NUMERICAL METHODS FOR SCIENTISTS AND ENGINEERS, Richard Hamming. Classic text stresses frequency approach in coverage of algorithms, polynomial approximation, Fourier approximation, exponential approximation, other topics. Revised and enlarged 2nd edition. 721pp. 5⅜ × 8½.
65241-6 Pa. $15.95

THEORETICAL SOLID STATE PHYSICS, Vol. I: Perfect Lattices in Equilibrium; Vol. II: Non-Equilibrium and Disorder, William Jones and Norman H. March. Monumental reference work covers fundamental theory of equilibrium properties of perfect crystalline solids, non-equilibrium properties, defects and disordered systems. Appendices. Problems. Preface. Diagrams. Index. Bibliography. Total of 1,301pp. 5⅜ × 8½. Two volumes.   Vol. I 65015-4 Pa. $16.95
Vol. II 65016-2 Pa. $14.95

OPTIMIZATION THEORY WITH APPLICATIONS, Donald A. Pierre. Broad-spectrum approach to important topic. Classical theory of minima and maxima, calculus of variations, simplex technique and linear programming, more. Many problems, examples. 640pp. 5⅜ × 8½.   65205-X Pa. $14.95

THE CONTINUUM: A Critical Examination of the Foundation of Analysis, Hermann Weyl. Classic of 20th-century foundational research deals with the conceptual problem posed by the continuum. 156pp. 5⅜ × 8½.   67982-9 Pa. $6.95

ESSAYS ON THE THEORY OF NUMBERS, Richard Dedekind. Two classic essays by great German mathematician: on the theory of irrational numbers; and on transfinite numbers and properties of natural numbers. 115pp. 5⅜ × 8½.
21010-3 Pa. $5.95

THE FUNCTIONS OF MATHEMATICAL PHYSICS, Harry Hochstadt. Comprehensive treatment of orthogonal polynomials, hypergeometric functions, Hill's equation, much more. Bibliography. Index. 322pp. 5⅜ × 8½.   65214-9 Pa. $9.95

NUMBER THEORY AND ITS HISTORY, Oystein Ore. Unusually clear, accessible introduction covers counting, properties of numbers, prime numbers, much more. Bibliography. 380pp. 5⅜ × 8½.   65620-9 Pa. $9.95

THE VARIATIONAL PRINCIPLES OF MECHANICS, Cornelius Lanczos. Graduate level coverage of calculus of variations, equations of motion, relativistic mechanics, more. First inexpensive paperbound edition of classic treatise. Index. Bibliography. 418pp. 5⅜ × 8½.   65067-7 Pa. $12.95

MATHEMATICAL TABLES AND FORMULAS, Robert D. Carmichael and Edwin R. Smith. Logarithms, sines, tangents, trig functions, powers, roots, reciprocals, exponential and hyperbolic functions, formulas and theorems. 269pp. 5⅜ × 8½.   60111-0 Pa. $6.95

THEORETICAL PHYSICS, Georg Joos, with Ira M. Freeman. Classic overview covers essential math, mechanics, electromagnetic theory, thermodynamics, quantum mechanics, nuclear physics, other topics. First paperback edition. xxiii + 885pp. 5⅜ × 8½.   65227-0 Pa. $21.95

HANDBOOK OF MATHEMATICAL FUNCTIONS WITH FORMULAS, GRAPHS, AND MATHEMATICAL TABLES, edited by Milton Abramowitz and Irene A. Stegun. Vast compendium: 29 sets of tables, some to as high as 20 places. 1,046pp. 8 × 10½. 61272-4 Pa. $24.95

MATHEMATICAL METHODS IN PHYSICS AND ENGINEERING, John W. Dettman. Algebraically based approach to vectors, mapping, diffraction, other topics in applied math. Also generalized functions, analytic function theory, more. Exercises. 448pp. 5⅜ × 8¼. 65649-7 Pa. $10.95

A SURVEY OF NUMERICAL MATHEMATICS, David M. Young and Robert Todd Gregory. Broad self-contained coverage of computer-oriented numerical algorithms for solving various types of mathematical problems in linear algebra, ordinary and partial, differential equations, much more. Exercises. Total of 1,248pp. 5⅜ × 8½. Two volumes. Vol. I 65691-8 Pa. $14.95
Vol. II 65692-6 Pa. $14.95

TENSOR ANALYSIS FOR PHYSICISTS, J.A. Schouten. Concise exposition of the mathematical basis of tensor analysis, integrated with well-chosen physical examples of the theory. Exercises. Index. Bibliography. 289pp. 5⅜ × 8½. 65582-2 Pa. $8.95

INTRODUCTION TO NUMERICAL ANALYSIS (2nd Edition), F.B. Hildebrand. Classic, fundamental treatment covers computation, approximation, interpolation, numerical differentiation and integration, other topics. 150 new problems. 669pp. 5⅜ × 8½. 65363-3 Pa. $15.95

INVESTIGATIONS ON THE THEORY OF THE BROWNIAN MOVEMENT, Albert Einstein. Five papers (1905–8) investigating dynamics of Brownian motion and evolving elementary theory. Notes by R. Fürth. 122pp. 5⅜ × 8½. 60304-0 Pa. $4.95

CATASTROPHE THEORY FOR SCIENTISTS AND ENGINEERS, Robert Gilmore. Advanced-level treatment describes mathematics of theory grounded in the work of Poincaré, R. Thom, other mathematicians. Also important applications to problems in mathematics, physics, chemistry and engineering. 1981 edition. References. 28 tables. 397 black-and-white illustrations. xvii + 666pp. 6⅛ × 9¼. 67539-4 Pa. $17.95

AN INTRODUCTION TO STATISTICAL THERMODYNAMICS, Terrell L. Hill. Excellent basic text offers wide-ranging coverage of quantum statistical mechanics, systems of interacting molecules, quantum statistics, more. 523pp. 5⅜ × 8½. 65242-4 Pa. $12.95

STATISTICAL PHYSICS, Gregory H. Wannier. Classic text combines thermodynamics, statistical mechanics and kinetic theory in one unified presentation of thermal physics. Problems with solutions. Bibliography. 532pp. 5⅜ × 8½. 65401-X Pa. $12.95

ORDINARY DIFFERENTIAL EQUATIONS, Morris Tenenbaum and Harry Pollard. Exhaustive survey of ordinary differential equations for undergraduates in mathematics, engineering, science. Thorough analysis of theorems. Diagrams. Bibliography. Index. 818pp. 5⅜ × 8½. 64940-7 Pa. $18.95

STATISTICAL MECHANICS: Principles and Applications, Terrell L. Hill. Standard text covers fundamentals of statistical mechanics, applications to fluctuation theory, imperfect gases, distribution functions, more. 448pp. 5⅜ × 8½. 65390-0 Pa. $11.95

ORDINARY DIFFERENTIAL EQUATIONS AND STABILITY THEORY: An Introduction, David A. Sánchez. Brief, modern treatment. Linear equation, stability theory for autonomous and nonautonomous systems, etc. 164pp. 5⅜ × 8¼. 63828-6 Pa. $6.95

THIRTY YEARS THAT SHOOK PHYSICS: The Story of Quantum Theory, George Gamow. Lucid, accessible introduction to influential theory of energy and matter. Careful explanations of Dirac's anti-particles, Bohr's model of the atom, much more. 12 plates. Numerous drawings. 240pp. 5⅜ × 8½. 24895-X Pa. $6.95

THEORY OF MATRICES, Sam Perlis. Outstanding text covering rank, nonsingularity and inverses in connection with the development of canonical matrices under the relation of equivalence, and without the intervention of determinants. Includes exercises. 237pp. 5⅜ × 8½. 66810-X Pa. $8.95

GREAT EXPERIMENTS IN PHYSICS: Firsthand Accounts from Galileo to Einstein, edited by Morris H. Shamos. 25 crucial discoveries: Newton's laws of motion, Chadwick's study of the neutron, Hertz on electromagnetic waves, more. Original accounts clearly annotated. 370pp. 5⅜ × 8½. 25346-5 Pa. $10.95

INTRODUCTION TO PARTIAL DIFFERENTIAL EQUATIONS WITH APPLICATIONS, E.C. Zachmanoglou and Dale W. Thoe. Essentials of partial differential equations applied to common problems in engineering and the physical sciences. Problems and answers. 416pp. 5⅜ × 8½. 65251-3 Pa. $11.95

BURNHAM'S CELESTIAL HANDBOOK, Robert Burnham, Jr. Thorough guide to the stars beyond our solar system. Exhaustive treatment. Alphabetical by constellation: Andromeda to Cetus in Vol. 1; Chamaeleon to Orion in Vol. 2; and Pavo to Vulpecula in Vol. 3. Hundreds of illustrations. Index in Vol. 3. 2,000pp. 6⅛ × 9¼. 23567-X, 23568-8, 23673-0 Pa., Three-vol. set $44.85

CHEMICAL MAGIC, Leonard A. Ford. Second Edition, Revised by E. Winston Grundmeier. Over 100 unusual stunts demonstrating cold fire, dust explosions, much more. Text explains scientific principles and stresses safety precautions. 128pp. 5⅜ × 8½. 67628-5 Pa. $5.95

AMATEUR ASTRONOMER'S HANDBOOK, J.B. Sidgwick. Timeless, comprehensive coverage of telescopes, mirrors, lenses, mountings, telescope drives, micrometers, spectroscopes, more. 189 illustrations. 576pp. 5⅜ × 8¼. (Available in U.S. only) 24034-7 Pa. $11.95

SPECIAL FUNCTIONS, N.N. Lebedev. Translated by Richard Silverman. Famous Russian work treating more important special functions, with applications to specific problems of physics and engineering. 38 figures. 308pp. 5⅜ × 8½.
60624-4 Pa. $9.95

OBSERVATIONAL ASTRONOMY FOR AMATEURS, J.B. Sidgwick. Mine of useful data for observation of sun, moon, planets, asteroids, aurorae, meteors, comets, variables, binaries, etc. 39 illustrations. 384pp. 5⅜ × 8¼. (Available in U.S. only)
24033-9 Pa. $8.95

INTEGRAL EQUATIONS, F.G. Tricomi. Authoritative, well-written treatment of extremely useful mathematical tool with wide applications. Volterra Equations, Fredholm Equations, much more. Advanced undergraduate to graduate level. Exercises. Bibliography. 238pp. 5⅜ × 8½.
64828-1 Pa. $8.95

POPULAR LECTURES ON MATHEMATICAL LOGIC, Hao Wang. Noted logician's lucid treatment of historical developments, set theory, model theory, recursion theory and constructivism, proof theory, more. 3 appendixes. Bibliography. 1981 edition. ix + 283pp. 5⅜ × 8½.
67632-3 Pa. $8.95

MODERN NONLINEAR EQUATIONS, Thomas L. Saaty. Emphasizes practical solution of problems; covers seven types of equations. ". . . a welcome contribution to the existing literature. . . ."—Math Reviews. 490pp. 5⅜ × 8½. 64232-1 Pa. $11.95

FUNDAMENTALS OF ASTRODYNAMICS, Roger Bate et al. Modern approach developed by U.S. Air Force Academy. Designed as a first course. Problems, exercises. Numerous illustrations. 455pp. 5⅜ × 8½.
60061-0 Pa. $9.95

INTRODUCTION TO LINEAR ALGEBRA AND DIFFERENTIAL EQUATIONS, John W. Dettman. Excellent text covers complex numbers, determinants, orthonormal bases, Laplace transforms, much more. Exercises with solutions. Undergraduate level. 416pp. 5⅜ × 8½.
65191-6 Pa. $10.95

INCOMPRESSIBLE AERODYNAMICS, edited by Bryan Thwaites. Covers theoretical and experimental treatment of the uniform flow of air and viscous fluids past two-dimensional aerofoils and three-dimensional wings; many other topics. 654pp. 5⅜ × 8½.
65465-6 Pa. $16.95

INTRODUCTION TO DIFFERENCE EQUATIONS, Samuel Goldberg. Exceptionally clear exposition of important discipline with applications to sociology, psychology, economics. Many illustrative examples; over 250 problems. 260pp. 5⅜ × 8½.
65084-7 Pa. $8.95

LAMINAR BOUNDARY LAYERS, edited by L. Rosenhead. Engineering classic covers steady boundary layers in two- and three-dimensional flow, unsteady boundary layers, stability, observational techniques, much more. 708pp. 5⅜ × 8½.
65646-2 Pa. $18.95

LECTURES ON CLASSICAL DIFFERENTIAL GEOMETRY, Second Edition, Dirk J. Struik. Excellent brief introduction covers curves, theory of surfaces, fundamental equations, geometry on a surface, conformal mapping, other topics. Problems. 240pp. 5⅜ × 8½.
65609-8 Pa. $8.95

ROTARY-WING AERODYNAMICS, W.Z. Stepniewski. Clear, concise text covers aerodynamic phenomena of the rotor and offers guidelines for helicopter performance evaluation. Originally prepared for NASA. 537 figures. 640pp. 6⅛ × 9¼.
64647-5 Pa. $15.95

DIFFERENTIAL GEOMETRY, Heinrich W. Guggenheimer. Local differential geometry as an application of advanced calculus and linear algebra. Curvature, transformation groups, surfaces, more. Exercises. 62 figures. 378pp. 5⅜ × 8½.
63433-7 Pa. $9.95

INTRODUCTION TO SPACE DYNAMICS, William Tyrrell Thomson. Comprehensive, classic introduction to space-flight engineering for advanced undergraduate and graduate students. Includes vector algebra, kinematics, transformation of coordinates. Bibliography. Index. 352pp. 5⅜ × 8½. 65113-4 Pa. $9.95

A SURVEY OF MINIMAL SURFACES, Robert Osserman. Up-to-date, in-depth discussion of the field for advanced students. Corrected and enlarged edition covers new developments. Includes numerous problems. 192pp. 5⅜ × 8½.
64998-9 Pa. $8.95

ANALYTICAL MECHANICS OF GEARS, Earle Buckingham. Indispensable reference for modern gear manufacture covers conjugate gear-tooth action, gear-tooth profiles of various gears, many other topics. 263 figures. 102 tables. 546pp. 5⅜ × 8½. 65712-4 Pa. $14.95

SET THEORY AND LOGIC, Robert R. Stoll. Lucid introduction to unified theory of mathematical concepts. Set theory and logic seen as tools for conceptual understanding of real number system. 496pp. 5⅜ × 8¼. 63829-4 Pa. $12.95

A HISTORY OF MECHANICS, René Dugas. Monumental study of mechanical principles from antiquity to quantum mechanics. Contributions of ancient Greeks, Galileo, Leonardo, Kepler, Lagrange, many others. 671pp. 5⅜ × 8½.
65632-2 Pa. $14.95

FAMOUS PROBLEMS OF GEOMETRY AND HOW TO SOLVE THEM, Benjamin Bold. Squaring the circle, trisecting the angle, duplicating the cube: learn their history, why they are impossible to solve, then solve them yourself. 128pp. 5⅜ × 8½. 24297-8 Pa. $4.95

MECHANICAL VIBRATIONS, J.P. Den Hartog. Classic textbook offers lucid explanations and illustrative models, applying theories of vibrations to a variety of practical industrial engineering problems. Numerous figures. 233 problems, solutions. Appendix. Index. Preface. 436pp. 5⅜ × 8½. 64785-4 Pa. $11.95

CURVATURE AND HOMOLOGY, Samuel I. Goldberg. Thorough treatment of specialized branch of differential geometry. Covers Riemannian manifolds, topology of differentiable manifolds, compact Lie groups, other topics. Exercises. 315pp. 5⅜ × 8½. 64314-X Pa. $9.95

HISTORY OF STRENGTH OF MATERIALS, Stephen P. Timoshenko. Excellent historical survey of the strength of materials with many references to the theories of elasticity and structure. 245 figures. 452pp. 5⅜ × 8½. 61187-6 Pa. $12.95

GEOMETRY OF COMPLEX NUMBERS, Hans Schwerdtfeger. Illuminating, widely praised book on analytic geometry of circles, the Moebius transformation, and two-dimensional non-Euclidean geometries. 200pp. 5⅜ × 8¼.
63830-8 Pa. $8.95

MECHANICS, J.P. Den Hartog. A classic introductory text or refresher. Hundreds of applications and design problems illuminate fundamentals of trusses, loaded beams and cables, etc. 334 answered problems. 462pp. 5⅜ × 8½. 60754-2 Pa. $10.95

TOPOLOGY, John G. Hocking and Gail S. Young. Superb one-year course in classical topology. Topological spaces and functions, point-set topology, much more. Examples and problems. Bibliography. Index. 384pp. 5⅜ × 8¼.
65676-4 Pa. $10.95

STRENGTH OF MATERIALS, J.P. Den Hartog. Full, clear treatment of basic material (tension, torsion, bending, etc.) plus advanced material on engineering methods, applications. 350 answered problems. 323pp. 5⅜ × 8½. 60755-0 Pa. $9.95

ELEMENTARY CONCEPTS OF TOPOLOGY, Paul Alexandroff. Elegant, intuitive approach to topology from set-theoretic topology to Betti groups; how concepts of topology are useful in math and physics. 25 figures. 57pp. 5⅜ × 8½.
60747-X Pa. $3.95

ADVANCED STRENGTH OF MATERIALS, J.P. Den Hartog. Superbly written advanced text covers torsion, rotating disks, membrane stresses in shells, much more. Many problems and answers. 388pp. 5⅜ × 8½. 65407-9 Pa. $10.95

COMPUTABILITY AND UNSOLVABILITY, Martin Davis. Classic graduate-level introduction to theory of computability, usually referred to as theory of recurrent functions. New preface and appendix. 288pp. 5⅜ × 8½. 61471-9 Pa. $8.95

GENERAL CHEMISTRY, Linus Pauling. Revised 3rd edition of classic first-year text by Nobel laureate. Atomic and molecular structure, quantum mechanics, statistical mechanics, thermodynamics correlated with descriptive chemistry. Problems. 992pp. 5⅜ × 8½. 65622-5 Pa. $19.95

AN INTRODUCTION TO MATRICES, SETS AND GROUPS FOR SCIENCE STUDENTS, G. Stephenson. Concise, readable text introduces sets, groups, and most importantly, matrices to undergraduate students of physics, chemistry, and engineering. Problems. 164pp. 5⅜ × 8½. 65077-4 Pa. $7.95

THE HISTORICAL BACKGROUND OF CHEMISTRY, Henry M. Leicester. Evolution of ideas, not individual biography. Concentrates on formulation of a coherent set of chemical laws. 260pp. 5⅜ × 8½. 61053-5 Pa. $7.95

THE PHILOSOPHY OF MATHEMATICS: An Introductory Essay, Stephan Körner. Surveys the views of Plato, Aristotle, Leibniz & Kant concerning propositions and theories of applied and pure mathematics. Introduction. Two appendices. Index. 198pp. 5⅜ × 8½. 25048-2 Pa. $8.95

THE DEVELOPMENT OF MODERN CHEMISTRY, Aaron J. Ihde. Authoritative history of chemistry from ancient Greek theory to 20th-century innovation. Covers major chemists and their discoveries. 209 illustrations. 14 tables. Bibliographies. Indices. Appendices. 851pp. 5⅜ × 8½. 64235-6 Pa. $18.95

DE RE METALLICA, Georgius Agricola. The famous Hoover translation of greatest treatise on technological chemistry, engineering, geology, mining of early modern times (1556). All 289 original woodcuts. 638pp. 6¾ × 11.
60006-8 Pa. $18.95

SOME THEORY OF SAMPLING, William Edwards Deming. Analysis of the problems, theory and design of sampling techniques for social scientists, industrial managers and others who find statistics increasingly important in their work. 61 tables. 90 figures. xvii + 602pp. 5⅜ × 8½.
64684-X Pa. $15.95

THE VARIOUS AND INGENIOUS MACHINES OF AGOSTINO RAMELLI: A Classic Sixteenth-Century Illustrated Treatise on Technology, Agostino Ramelli. One of the most widely known and copied works on machinery in the 16th century. 194 detailed plates of water pumps, grain mills, cranes, more. 608pp. 9 × 12.
28180-9 Pa. $24.95

LINEAR PROGRAMMING AND ECONOMIC ANALYSIS, Robert Dorfman, Paul A. Samuelson and Robert M. Solow. First comprehensive treatment of linear programming in standard economic analysis. Game theory, modern welfare economics, Leontief input-output, more. 525pp. 5⅜ × 8½.
65491-5 Pa. $14.95

ELEMENTARY DECISION THEORY, Herman Chernoff and Lincoln E. Moses. Clear introduction to statistics and statistical theory covers data processing, probability and random variables, testing hypotheses, much more. Exercises. 364pp. 5⅜ × 8½.
65218-1 Pa. $10.95

THE COMPLEAT STRATEGYST: Being a Primer on the Theory of Games of Strategy, J.D. Williams. Highly entertaining classic describes, with many illustrated examples, how to select best strategies in conflict situations. Prefaces. Appendices. 268pp. 5⅜ × 8½.
25101-2 Pa. $7.95

CONSTRUCTIONS AND COMBINATORIAL PROBLEMS IN DESIGN OF EXPERIMENTS, Damaraju Raghavarao. In-depth reference work examines orthogonal Latin squares, incomplete block designs, tactical configuration, partial geometry, much more. Abundant explanations, examples. 416pp. 5⅜ × 8¼.
65685-3 Pa. $10.95

THE ABSOLUTE DIFFERENTIAL CALCULUS (CALCULUS OF TENSORS), Tullio Levi-Civita. Great 20th-century mathematician's classic work on material necessary for mathematical grasp of theory of relativity. 452pp. 5⅜ × 8½.
63401-9 Pa. $11.95

VECTOR AND TENSOR ANALYSIS WITH APPLICATIONS, A.I. Borisenko and I.E. Tarapov. Concise introduction. Worked-out problems, solutions, exercises. 257pp. 5⅜ × 8¼.
63833-2 Pa. $8.95

THE FOUR-COLOR PROBLEM: Assaults and Conquest, Thomas L. Saaty and Paul G. Kainen. Engrossing, comprehensive account of the century-old combinatorial topological problem, its history and solution. Bibliographies. Index. 110 figures. 228pp. 5⅜ × 8½.   65092-8 Pa. $6.95

CATALYSIS IN CHEMISTRY AND ENZYMOLOGY, William P. Jencks. Exceptionally clear coverage of mechanisms for catalysis, forces in aqueous solution, carbonyl- and acyl-group reactions, practical kinetics, more. 864pp. 5⅜ × 8½.   65460-5 Pa. $19.95

PROBABILITY: An Introduction, Samuel Goldberg. Excellent basic text covers set theory, probability theory for finite sample spaces, binomial theorem, much more. 360 problems. Bibliographies. 322pp. 5⅜ × 8½.   65252-1 Pa. $9.95

LIGHTNING, Martin A. Uman. Revised, updated edition of classic work on the physics of lightning. Phenomena, terminology, measurement, photography, spectroscopy, thunder, more. Reviews recent research. Bibliography. Indices. 320pp. 5⅜ × 8¼.   64575-4 Pa. $8.95

PROBABILITY THEORY: A Concise Course, Y.A. Rozanov. Highly readable, self-contained introduction covers combination of events, dependent events, Bernoulli trials, etc. Translation by Richard Silverman. 148pp. 5⅜ × 8¼.
63544-9 Pa. $6.95

AN INTRODUCTION TO HAMILTONIAN OPTICS, H. A. Buchdahl. Detailed account of the Hamiltonian treatment of aberration theory in geometrical optics. Many classes of optical systems defined in terms of the symmetries they possess. Problems with detailed solutions. 1970 edition. xv + 360pp. 5⅜ × 8½.
67597-1 Pa. $10.95

STATISTICS MANUAL, Edwin L. Crow, et al. Comprehensive, practical collection of classical and modern methods prepared by U.S. Naval Ordnance Test Station. Stress on use. Basics of statistics assumed. 288pp. 5⅜ × 8½.
60599-X Pa. $7.95

DICTIONARY/OUTLINE OF BASIC STATISTICS, John E. Freund and Frank J. Williams. A clear concise dictionary of over 1,000 statistical terms and an outline of statistical formulas covering probability, nonparametric tests, much more. 208pp. 5⅜ × 8½.   66796-0 Pa. $7.95

STATISTICAL METHOD FROM THE VIEWPOINT OF QUALITY CONTROL, Walter A. Shewhart. Important text explains regulation of variables, uses of statistical control to achieve quality control in industry, agriculture, other areas. 192pp. 5⅜ × 8½.   65232-7 Pa. $7.95

THE INTERPRETATION OF GEOLOGICAL PHASE DIAGRAMS, Ernest G. Ehlers. Clear, concise text emphasizes diagrams of systems under fluid or containing pressure; also coverage of complex binary systems, hydrothermal melting, more. 288pp. 6½ × 9¼.   65389-7 Pa. $10.95

STATISTICAL ADJUSTMENT OF DATA, W. Edwards Deming. Introduction to basic concepts of statistics, curve fitting, least squares solution, conditions without parameter, conditions containing parameters. 26 exercises worked out. 271pp. 5⅜ × 8½.   64685-8 Pa. $9.95

TENSOR CALCULUS, J.L. Synge and A. Schild. Widely used introductory text covers spaces and tensors, basic operations in Riemannian space, non-Riemannian spaces, etc. 324pp. 5⅜ × 8¼. 63612-7 Pa. $9.95

A CONCISE HISTORY OF MATHEMATICS, Dirk J. Struik. The best brief history of mathematics. Stresses origins and covers every major figure from ancient Near East to 19th century. 41 illustrations. 195pp. 5⅜ × 8½. 60255-9 Pa. $7.95

A SHORT ACCOUNT OF THE HISTORY OF MATHEMATICS, W.W. Rouse Ball. One of clearest, most authoritative surveys from the Egyptians and Phoenicians through 19th-century figures such as Grassman, Galois, Riemann. Fourth edition. 522pp. 5⅜ × 8½. 20630-0 Pa. $11.95

HISTORY OF MATHEMATICS, David E. Smith. Nontechnical survey from ancient Greece and Orient to late 19th century; evolution of arithmetic, geometry, trigonometry, calculating devices, algebra, the calculus. 362 illustrations. 1,355pp. 5⅜ × 8½. 20429-4, 20430-8 Pa., Two-vol. set $26.90

THE GEOMETRY OF RENÉ DESCARTES, René Descartes. The great work founded analytical geometry. Original French text, Descartes' own diagrams, together with definitive Smith-Latham translation. 244pp. 5⅜ × 8½.
60068-8 Pa. $7.95

THE ORIGINS OF THE INFINITESIMAL CALCULUS, Margaret E. Baron. Only fully detailed and documented account of crucial discipline: origins; development by Galileo, Kepler, Cavalieri; contributions of Newton, Leibniz, more. 304pp. 5⅜ × 8½. (Available in U.S. and Canada only) 65371-4 Pa. $9.95

THE HISTORY OF THE CALCULUS AND ITS CONCEPTUAL DEVELOPMENT, Carl B. Boyer. Origins in antiquity, medieval contributions, work of Newton, Leibniz, rigorous formulation. Treatment is verbal. 346pp. 5⅜ × 8½.
60509-4 Pa. $9.95

THE THIRTEEN BOOKS OF EUCLID'S ELEMENTS, translated with introduction and commentary by Sir Thomas L. Heath. Definitive edition. Textual and linguistic notes, mathematical analysis. 2,500 years of critical commentary. Not abridged. 1,414pp. 5⅜ × 8½. 60088-2, 60089-0, 60090-4 Pa., Three-vol. set $31.85

GAMES AND DECISIONS: Introduction and Critical Survey, R. Duncan Luce and Howard Raiffa. Superb nontechnical introduction to game theory, primarily applied to social sciences. Utility theory, zero-sum games, n-person games, decision-making, much more. Bibliography. 509pp. 5⅜ × 8½. 65943-7 Pa. $12.95

THE HISTORICAL ROOTS OF ELEMENTARY MATHEMATICS, Lucas N.H. Bunt, Phillip S. Jones, and Jack D. Bedient. Fundamental underpinnings of modern arithmetic, algebra, geometry and number systems derived from ancient civilizations. 320pp. 5⅜ × 8½. 25563-8 Pa. $8.95

CALCULUS REFRESHER FOR TECHNICAL PEOPLE, A. Albert Klaf. Covers important aspects of integral and differential calculus via 756 questions. 566 problems, most answered. 431pp. 5⅜ × 8½. 20370-0 Pa. $8.95

CHALLENGING MATHEMATICAL PROBLEMS WITH ELEMENTARY SOLUTIONS, A.M. Yaglom and I.M. Yaglom. Over 170 challenging problems on probability theory, combinatorial analysis, points and lines, topology, convex polygons, many other topics. Solutions. Total of 445pp. 5⅜ × 8½. Two-vol. set.
Vol. I 65536-9 Pa. $7.95
Vol. II 65537-7 Pa. $7.95

FIFTY CHALLENGING PROBLEMS IN PROBABILITY WITH SOLUTIONS, Frederick Mosteller. Remarkable puzzlers, graded in difficulty, illustrate elementary and advanced aspects of probability. Detailed solutions. 88pp. 5⅜ × 8½.
65355-2 Pa. $4.95

EXPERIMENTS IN TOPOLOGY, Stephen Barr. Classic, lively explanation of one of the byways of mathematics. Klein bottles, Moebius strips, projective planes, map coloring, problem of the Koenigsberg bridges, much more, described with clarity and wit. 43 figures. 210pp. 5⅜ × 8½.
25933-1 Pa. $6.95

RELATIVITY IN ILLUSTRATIONS, Jacob T. Schwartz. Clear nontechnical treatment makes relativity more accessible than ever before. Over 60 drawings illustrate concepts more clearly than text alone. Only high school geometry needed. Bibliography. 128pp. 6⅛ × 9¼.
25965-X Pa. $7.95

AN INTRODUCTION TO ORDINARY DIFFERENTIAL EQUATIONS, Earl A. Coddington. A thorough and systematic first course in elementary differential equations for undergraduates in mathematics and science, with many exercises and problems (with answers). Index. 304pp. 5⅜ × 8½.
65942-9 Pa. $8.95

FOURIER SERIES AND ORTHOGONAL FUNCTIONS, Harry F. Davis. An incisive text combining theory and practical example to introduce Fourier series, orthogonal functions and applications of the Fourier method to boundary-value problems. 570 exercises. Answers and notes. 416pp. 5⅜ × 8½.
65973-9 Pa. $11.95

AN INTRODUCTION TO ALGEBRAIC STRUCTURES, Joseph Landin. Superb self-contained text covers "abstract algebra": sets and numbers, theory of groups, theory of rings, much more. Numerous well-chosen examples, exercises. 247pp. 5⅜ × 8½.
65940-2 Pa. $8.95

Prices subject to change without notice.
Available at your book dealer or write for free Mathematics and Science Catalog to Dept. GI, Dover Publications, Inc., 31 East 2nd St., Mineola, N.Y. 11501. Dover publishes more than 175 books each year on science, elementary and advanced mathematics, biology, music, art, literature, history, social sciences and other areas.